Kubiczki
W9-BNQ-376

Design of REMEDIATION SYSTEMS

Jimmy H.C. Wong
Chin Hong Lim
Greg L. Nolen

Acquiring Editor:	Joel Stein
Project Editor:	Joan Moscrop
Marketing Manager:	Greg Daurelle
Direct Marketing Manager:	Arline Massey
Cover design:	Denise Craig
Manufacturing:	Sheri Schwartz

Library of Congress Cataloging-in-Publication Data

Wong, Jimmy H. C.
Design of remediation systems / Jimmy H. C. Wong, Chin Hong Lim, Greg L. Nolen.
p. cm.
Includes bibliographical references and index.
ISBN 1-56670-217-8
1. Oil pollution of soils. 2. Oil pollution of water. 3. Hazardous waste site remediation.
I. Lim, Chin. II. Nolan, Greg. III. Title.
TD879.P4W65 1997
628.5′2—dc20 96-43810
CIP

International Standard Book Number 1-56670-217-8
Library of Congress Card Number 96-43810
Printed in the United States of America 1 2 3 4 5 6 7 8 9 0
Printed on acid-free paper

To my mother and late father
Jimmy H. C. Wong

To my son Darrin and my wife Siew Ping
Chin Hong Lim

To my adoring wife Julie, for providing me ample motivation
Greg L. Nolen

PREFACE

While petroleum-contained sites were recognized as a major threat to the environment in the early 1980s, the cleanup of these sites did not take off on a large scale until the late 1980s. During that time, states rushed to push regulations on the removal of underground storage tanks, and implemented procedures for conducting remedial investigations. These translated into a proliferation in cleanup technology and equipment.

During the initial "learning" stage, crude rule-of-thumb procedures were developed — more often than not, from trial and error attempts. Coupled with greater research at both the academic, owner/operator (e.g., petroleum companies) levels, this experience evolved to the present stage. Today, soil vapor extraction, sparging, bioremediation, pump-and-treat, and perhaps, to a lesser extent, intrinsic remediation, are considered the treatment methods of choice.

This book was written to provide a guide for environmental consultants, managers, owners and operators, regulators, and students to perform a remediation design from the assessment phase to completion. It strives to provide engineers with the tools to conduct a pilot test, apply the results, and design a system that is practical and efficient. It utilizes the authors' diversified and combined experience, including design and construction. The authors have long felt that a book addressing pure design issues was needed. It is their sincere hope that this endeavor fills that void.

AUTHORS

Jimmy H. C. Wong is the President of International Environmental Services, a consulting firm in Sacramento, California. He was previously affiliated with other consulting firms both in the Midwest and California. Throughout his professional career, Mr. Wong has been extensively involved in a wide variety of projects involving cleanup of subsurface contaminants. His experience in remediation technologies includes vapor extraction, sparging, bioremediation, and pump and treat.

Mr. Wong received his BS and MS in Civil Engineering from South Dakota State University. He is on the ASTM ISO 14000 Technical Committee and the Water Environment Federation's Asia/Pacific Rim Steering Committee. Mr. Wong is a registered professional engineer.

Chin Hong Lim is a Chemical/Environmental Engineer with HWS Consulting Group, Inc. in Lincoln, Nebraska. He has designed numerous groundwater and soil remediation systems throughout the Midwest. He is the primary design engineer who designs and specifies equipment for groundwater and soil remediation systems. Mr. Lim holds BS and MS degrees in Chemical Engineering from the University of Nebraska-Lincoln. He is currently pursuing a PhD degree in Chemical Engineering at the University of Nebraska-Lincoln. Mr. Lim is a member of American Institute of Chemical Engineers, National Society of Professional Engineers, American Water Works Association, and International Association on Water Quality.

Greg L. Nolen is the Manager of construction activities for the hazardous waste division of the Sacramento, California office of EMCON. He performs engineering design and constructability design reviews on projects ranging from potable water to land fill gas migration control. The majority of his projects involve groundwater remediation and off-site migration control. He has designed and constructed automated systems for major petroleum and utility companies and recently designed and supervised construction of a Superfund groundwater remediation project. Mr. Nolen has a certificate degree in construction management from the University of California, Davis and is a licensed California contractor.

CONTENTS

CHAPTER 6

1 INTRODUCTION

This book aims to be a comprehensive design manual for the design of remediation systems for the treatment of soils and groundwater contaminated with petroleum hydrocarbons. This manual was written for professionals familiar with remediation concepts, the design process, and the regulatory basis of cleanup control, as well as less experienced designers in the field of remediation engineering.

This book focuses on remediation practices from the design engineer's point of view. Its overriding theme is efficient and reliable design based on sound engineering principles and practical construction considerations. The book begins with discussions of concepts and principles and concludes with detailed engineering design processes. Real-life design examples are provided at the end of most chapters to give the reader a feel for actual applications of the concepts and processes described.

The book is divided into eight chapters, including this introduction to the book. The materials presented in this book were selected to provide the reader with an adequate single-source manual for the design of remediation systems. Chapter 2 presents a refresher course on the chemistry of hydrocarbons, including useful tables on various properties of petroleum chemicals. Chemical properties are discussed only to the extent that they affect the treatment of the contaminants. This chapter also describes the various laboratory analytical tests available and the distinctions between each of the tests.

Chapter 3 reviews geology and hydrogeology concepts that are essential from a design engineer's perspective. It discusses soil classifications, hydrocarbon phases in soils and groundwater, geology and hydrology concepts, groundwater flow fundamentals, and contaminant transport phenomena.

Chapter 4 evaluates the general design approach for remediation projects. It takes the reader step-by-step through a typical design process. In light of the current regulatory environment, a section on designs based on risk-based corrective action is included. Chapter 5 discusses soil vapor extraction and air sparging technology. Concepts and important design parameters associated with vapor extraction and sparge technologies are presented in a concise fashion. This is followed with discussions on procedures related to conducting vapor extraction and air sparging pilot tests. Detailed design and construction considerations are discussed.

Chapter 6 addresses the application of bioremediation in the treatment of contaminated soils and groundwater. The microbiology of bioremediation and the factors and conditions affecting the biodegradation process are discussed. Feasibility study and detailed design procedures are then presented, followed with a section on the implementation of intrinsic bioremediation. Chapter 7 covers pump and treat systems for remediating contaminated aquifers. The chapter focuses on well designs, pump selection, and the design of air stripping towers and activated carbon units. Chapter 8 presents a brief overview of innovative technologies for the treatment of contaminated soils and groundwater.

The design concepts and approach presented in this book are applicable to the majority of cleanup sites regardless of contaminant type or site location. Because of the voluminous material related to this subject, the book deals only with the remediation of petroleum products, principally the lighter hydrocarbons. The intent of this book is to provide environmental consultants, managers, owners and operators, and college students with the tools necessary to perform a remediation design from the assessment phase to completion.

The design of remediation systems is an imperfect science at best, which is compounded by the unseen problem found below ground. As such, there are no hard and fast rules dictating successful design. However, success can be measured by several yardsticks. From the regulator's point of view, treatment to cleanup levels would be one form of evaluation. Owners and operators can certainly be expected to demand cost-effective solutions, emphasizing low-cost systems with short cleanup times. The contractor's perspective would take into consideration ease of construction and system understanding. It is the ultimate objective of this book to provide the designer with the basic tools to incorporate these parameters with sound engineering know-how and judgment into a successful design.

2 CHEMISTRY OF HYDROCARBONS

2.1 INTRODUCTION

Organic chemicals are widely used in all facets of industries including machine and electronic manufacturing, metal finishing, automotive and engine repair, dry cleaning, asphalt operations, dye manufacturing, agricultural activities, and food processing. The usage of organic chemicals is of increasing concern to regulators and the industry in general because of the contamination of soil and groundwater resulting from the mishandling and disposal of these chemicals. Typically, these organic chemicals are stored in drums and tanks. Many industries, specifically gasoline fueling stations and trucking companies, store their chemicals in underground storage tanks (UST). The presence of large quantities of chemicals, gasoline, and diesel fuel on-site is often an indicator of the potential for soil and groundwater contamination.

Most of the organic chemicals used in the industry contain halogenated and aromatic hydrocarbons. These compounds are the most common culprits of soil and groundwater contamination due to leaking UST or surface spills. The first step toward implementing a remediation program is to provide for a better understanding of the physical properties of the organic chemicals themselves. This chapter examines the physical properties, structure, and nomenclature of selected organic compounds. In addition, transformations of the organic chemicals via chemical reactions will also be discussed.

2.2 BASIC CHEMISTRY OF HYDROCARBONS AND FUNCTIONAL GROUPS

The specific organic compounds addressed in this book are the petroleum hydrocarbons and organic functional groups, especially halogenated compounds such as carbon tetrachloride, trichloroethylene, and perchloroethylene. The majority of sites an engineer has to deal with will probably involve one of these contaminants, with gasoline being the most common at the smaller sites.

2.2.1 Definitions

The majority of carbon-containing compounds are classified as organic compounds. All organic compounds contain carbon; virtually all contain hydrogen and possess at least one C–H bond in their chemical structure. Hydrocarbons form the most basic organic compound group; they contain only the elements hydrogen and carbon.

Hydrocarbons are divided into two classes, aliphatic hydrocarbon, which is a straight-chain hydrocarbon without a benzene ring, and aromatic hydrocarbon, which contains a benzene ring. A benzene ring contains six carbons joined in a ring structure with alternative single and double bonds. The formula for benzene is C_6H_6.

2.2.2 Aliphatic Hydrocarbons

The aliphatic hydrocarbons are divided into three groups based on how their carbon atoms are joined: alkanes (joined by single bonds), alkenes (joined by double bonds), and alkynes (joined by triple bonds). If there is a combination of bonds in an aliphatic hydrocarbon, the compound is classified based on the multiple bond.

2.2.2.1 Alkanes

Alkanes are also known as paraffins, or saturated hydrocarbons. Alkanes are characterized by having only one carbon–carbon (C–C) bond and being rather unreactive chemically. There are three types of alkanes: straight-chain, branched-chain, and cycloalkanes. The branched-chain hydrocarbons are the results of isomers, the hydrocarbons that have the same formula but different physical and chemical properties because they differ in the arrangement of the atoms in their molecules. The general formula for straight- and branched-chained alkanes is C_nH_{2n+2}, and for cycloalkanes is C_nH_{2n}. Examples of alkanes are shown as follows:

Type of Alkane	Molecular Formula	Name	Structural Formula
Straight-chain	CH_4	Methane	H | H—C—H | H
Straight-chain	C_5H_{12}	Pentane	H H H H H | | | | | H–C–C–C–C–C–H | | | | | H H H H H
Branched-chain	C_5H_{12}	2-Methylbutane	H | H–C–H H | H H | | | | H–C–C–C–C–H | | | | H H H H

Type of Alkane	Molecular Formula	Name	Structural Formula
Cycloalkane	C_4H_8	Cyclobutane	H H / H−C−C−H / H−C−C−H / H H

The organic chemicals are named according to the rules set by the International Union of Pure and Applied Chemistry (IUPAC). For alkanes these rules are as follow:

1. The base name of a compound is the name of the longest straight-chain alkane.
2. Any chains that branch from the straight chain or other functional groups (if any) are called alkyl groups. The following are some examples for the alkyl groups:

Methyl	CH_3 —
Ethyl	CH_3CH_2 —
Propyl	$CH_3CH_2CH_2$ —
Isopropyl	CH_3CHCH_3 (with bond from central C)
Butyl	$CH_3CH_2CH_2CH_2$ —
t-Butyl	CH_3CCH_3 with CH_3 above central C (and bond below)

   Other functional groups include chloride ions, bromide ions, fluoride ions, indicated by the prefix chloro-, bromo-, and fluoro-, respectively.
3. The location of the functional group is indicated by numbering the carbon atoms in the longest straight chain, with the end carbon closest to the position of the first functional group being carbon number 1.
4. If there is more than one of the same functional group, the prefixes di-, tri-, and tetra- are used.
5. Consider the following example:

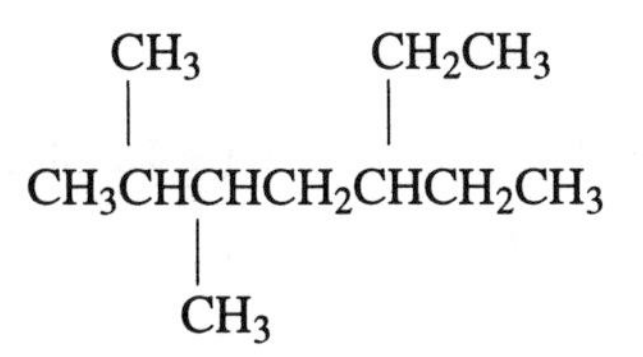

The longest straight chain has seven carbons, so this is a heptane. The carbons are numbered starting with the end closest to the position of an attached functional group. A methyl group is attached to the second and third carbons while an ethyl group is attached to the fifth carbon. Thus, the name of the compound is 2,3-dimethyl-5-ethylheptane.

2.2.2.2 *Alkenes*

Alkenes also called olefins or unsaturated hydrocarbons. They are characterized by carbon–carbon double bonds (C=C) and are named by locating the longest chain that contains the double bond. The base name of alkenes ends in *-ene* or *-ylene*. The general formula for alkene is C_nH_{2n}. The simplest and most widely manufactured alkene is ethylene, C_2H_4. Ethylene is a colorless gas with a sweet odor and, when mixed with oxygen, it is used as an anesthetic in dentistry and surgery. It has the property of causing green fruits to ripen and is used commercially for this purpose.

If the double bond can appear in more than one location, the location of the double bond is denoted by a numerical prefix, which identifies the carbon that contains the double bond. For example:

$CH_2{=}CHCH_2CH_2CH_3$ 1-Pentene

$CH_2CH{=}CHCH_2CH_3$ 2-Pentene

$CH_2CH_2CH{=}CHCH_3$ 3-Pentene

The position of any functional groups that are present is denoted by the carbon atom to which they are bonded, such as:

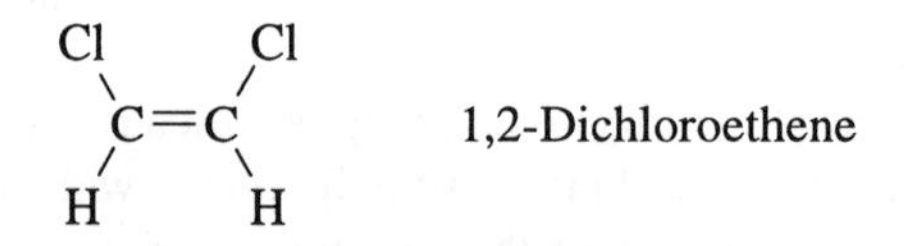

For alkanes, the carbon atoms can rotate along the single bonds so that any structural forms of an alkane are equivalent. With alkene, because of the double bond that prevents the rotation of carbon atoms, some alkenes exist as isomers. For example:

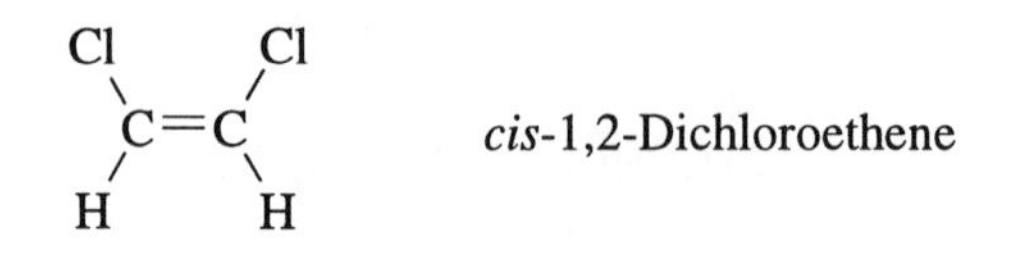

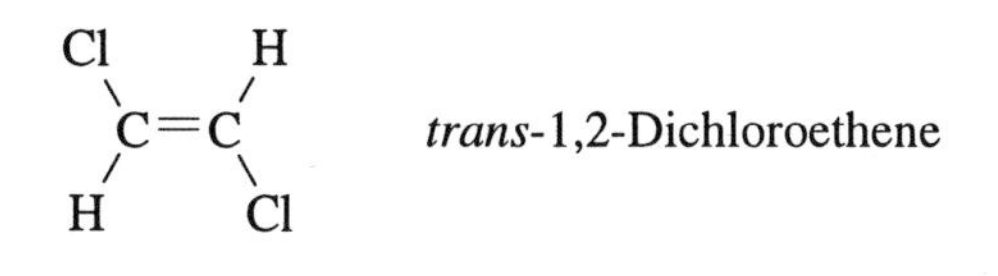

trans-1,2-Dichloroethene

2.2.2.3 Alkynes

Alkynes or acetylenic hydrocarbons contained triple carbon–carbon bonds (C≡C). The general formula for alkynes is C_nH_{2n-2}. Acetylene, C_2H_2, is an alkyne that is widely used as a welding gas. Acetylene and all the other alkynes burn very easily because the triple bond is a high energy bond, and this energy is released when the compound is transformed into carbon and water. Thus, the flame produced by burning acetylene is very hot and is suitable for welding and cutting metal.

H−C≡C−H   Acetylene

2.2.3 Aromatic Hydrocarbons

Aromatic hydrocarbons, also called alkyl hydrocarbons, are characterized by carbon–carbon ring structures, and, as suggested by their name, a distinctive odor. Benzene, C_6H_6, is the simplest compound of this group of hydrocarbons. Other molecules can be formed by joining functional groups to the basic benzene ring. If there is only one functional group attached to the benzene ring, the position of the functional group is not specified because all six carbon atoms are equivalent. If there are two functional groups attached, the new molecule will have three isomers. The naming of the three isomers is based on the locations of the two functional groups. For example:

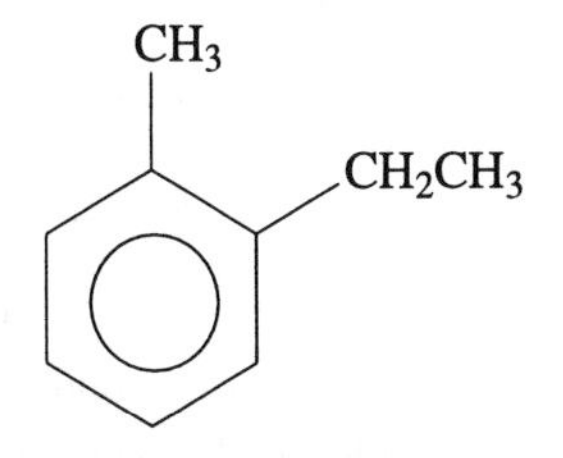

1-methyl-2-ethylbenzene

If the two functional groups attached to the benzene ring are the same, then the three isomers are named by attaching prefixes *ortho-*, *meta-*, and *para-*. For example:

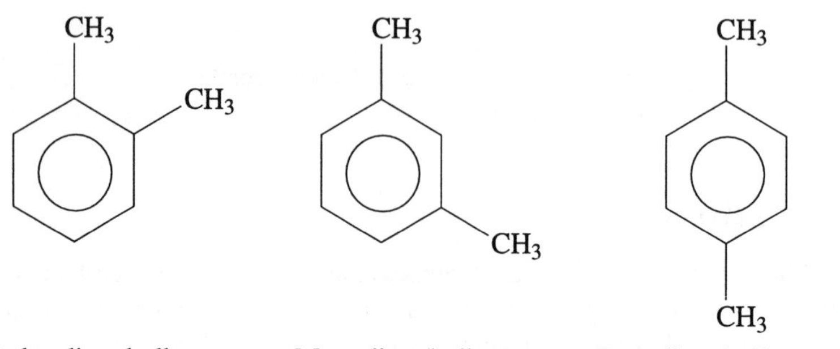

Ortho-dimethylbenzene Meta-dimethylbenzene Para-dimethylbenzene

If there are more than two functional groups attached to the benzene ring, the new molecule is named based on the location of the functional groups. For example:

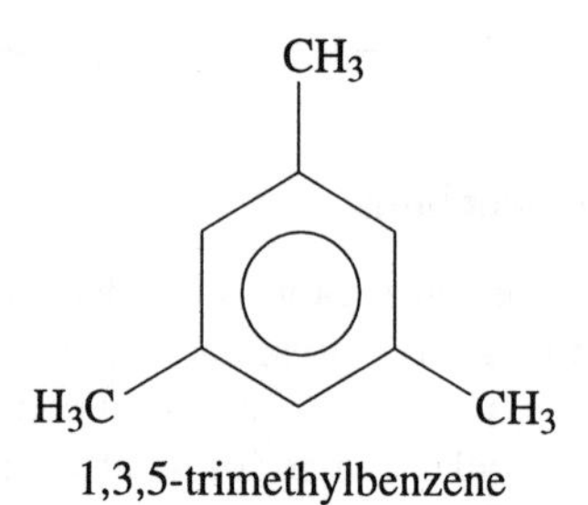

1,3,5-trimethylbenzene

The benzene ring itself can also be a functional group. In this case, it is called a phenyl- functional group. The most famous example of this case is polychlorinated biphenyl (PCB) where the biphenyl is chlorinated with 1 to 10 chlorine ions to yield a mixture of isomers.

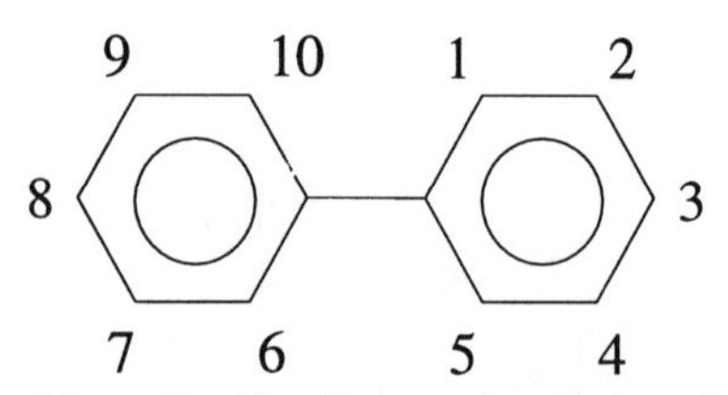

Possibility of 1 to 10 chloride ions bonded to the biphenyl.
Polychlorinated Biphenyl

2.2.3.1 *Alkylbenzenes*

Alkylbenzenes form an extensive family of aromatic hydrocarbons. They are formed by replacing the hydrogen in benzene with alkyl groups or halogens, such as chlorine. Alkyl groups refer to alkanes minus one hydrogen atom, as shown in the examples below.

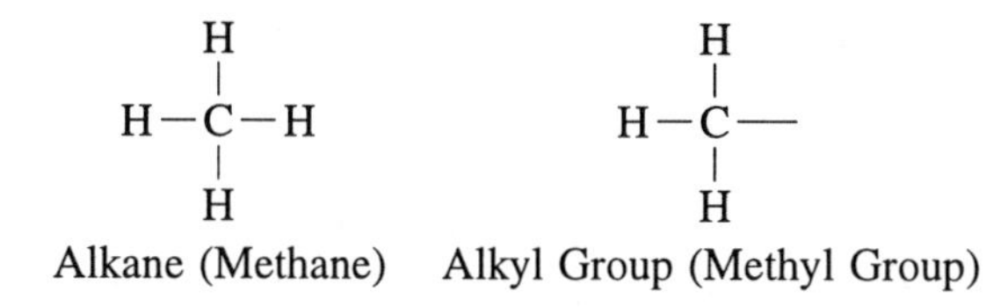

Alkane (Methane) Alkyl Group (Methyl Group)

1-methyl-2-ethylbenzene, ortho-dimethylbenzene, meta-dimethylbenzene, para-dimethylbenzene, and 1,3,5-trimethylbenzene are all examples of alkylbenzenes.

2.2.3.2 Polycyclic Aromatic Hydrocarbons

Polycyclic aromatic hydrocarbons (PAH) is another group of aromatic hydrocarbons that are characterized by multiple fused rings. This is often referred to as "chicken wire" structures. They are also called polynuclear aromatic hydrocarbons (PNA). Naphthalene ($C_{10}H_8$) and benzo(a)pyrene ($C_{20}H_{12}$) are two examples of a PAH.

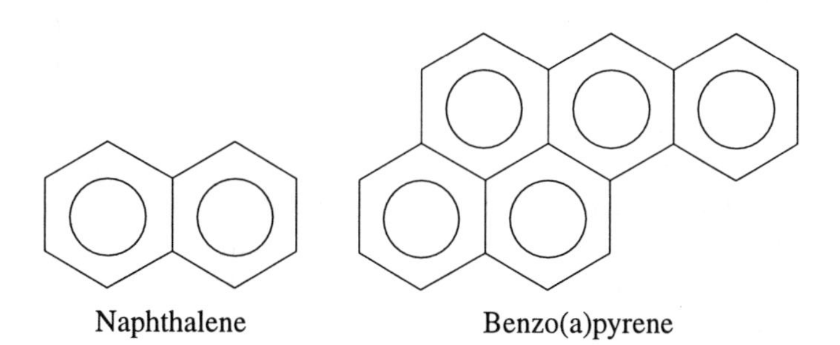
Naphthalene Benzo(a)pyrene

Table 2.1 lists the physical properties of some aromatic hydrocarbons that are commonly encountered in groundwater contamination (U.S. EPA, 1990b).

Table 2.1 Physical Properties for Selected Aromatic Hydrocarbons

Aromatic Hydrocarbons	Water Solubility (mg/L)	Vapor Pressure (mmHg)	Henry's Constant (atm-m^3/mol)
Benzene	1.75E+03	9.52E+01	5.59E–03
Benzo(a)anthracene	5.70E–03	2.20E–08	1.16E–06
Benzo(a)pyrene	1.20E–03	5.60E–09	1.55E–06
Bromobenzene	4.46E+02	4.14E+00	1.92E–03
Chlorobenzene	4.66E+02	1.17E+01	3.72E–03
Chlorotoluene	3.30E+03	1.00E+00	5.06E–05
1,2-Dichlorobenzene	1.00E+02	1.00E+00	1.93E–03
1,3-Dichlorobenzene	1.23E+02	2.28E+00	3.59E–03
1,4-Dichlorobenzene	7.90E+01	1.18E+00	2.89E–03

Table 2.1 Physical Properties for Selected Aromatic Hydrocarbons (Continued)

Aromatic Hydrocarbons	Water Solubility (mg/L)	Vapor Pressure (mmHg)	Henry's Constant (atm-m^3/mol)
2,4-Dichlorophenol	4.60E+03	5.90E–02	2.75E–06
Dichlorotoluene	2.50E+00	3.00E–01	2.54E–02
2,4-Dimethylphenol	4.20E+03	6.21E–02	2.38E–06
2,4-Dinitrophenol	5.60E+03	1.49E–05	6.45E-10
2,4-Dinitrotoluene	2.40E+02	5.10E–03	5.09E–06
2,6-Dinitrotoluene	1.32E+03	1.80E–02	3.27E–06
Ethylbenzene	1.52E+02	7.00E+00	6.43E–03
Hexachlorobenzene	6.00E–03	1.09E–05	6.81E–04
Naphthalene	3.17E+01	2.30E–01	1.15E–03
Nitrobenzene	1.90E+03	1.50E–01	2.20E–05
Pentachloronitrobenzene	7.11E–02	1.13E–04	6.18E–04
Pentachlorophenol	1.40E+01	1.10E–04	2.75E–06
Phenol	9.30E+04	3.41E–01	4.54E–07
Pyrene	1.32E–01	2.50E–06	5.04E–06
Styrene	3.00E+02	4.50E+00	2.05E–03
Toluene	5.35E+02	2.81E+01	6.37E–03
1,2,3-Trichlorobenzene	1.20E+01	2.10E–01	4.23E–03
1,2,4-Trichlorobenzene	3.00E+01	2.90E–01	2.31E–03
1,3,5-Trichlorobenzene	5.80E+00	5.80E–01	2.39E–02
2,4,5-Trichlorobenzene	1.19E+03	1.00E+00	2.18E–04
2,4,6-Trichlorobenzene	8.00E+02	1.20E–02	3.90E–06
1,2,4-Trimethylbenzene	5.76E+01	2.03E+00	5.57E–03
Total Xylenes	1.98E+02	1.00E+01	7.04E–03
m-Xylene	1.30E+02	1.00E+01	1.07E–02
o-Xylene	1.75E+02	6.06E+03	5.10E–03
p-Xylene	1.98E+02	1.00E+01	7.05E–03

From Johnson et al., *Groundwater Monitoring Review*, 1990. With permission.

2.2.4 Functional Groups

Functional groups, in general, contain at least one atom other than carbon or hydrogen. However, alkenes and alkynes by virtue of their double and triple bonds are also considered functional groups. The functional group organooxygen compounds refer to oxygen-containing compounds. Examples include alcohols, aldehydes, ethers, ketones, esters, carboxylic acids, and phenols. The functional group organohalide compounds refers to halogen-containing compounds. They contain at least one atom of F, Cl, Br, or I. There are two common types of organohalide compounds: alkyl halides and alkenyl halides.

Alkyl halides are formed by the substitution of halogen atoms for one or more hydrogen atoms on alkanes. Examples include carbon tetrachloride (CCl_4) and dichloromethane (CH_2Cl_2).

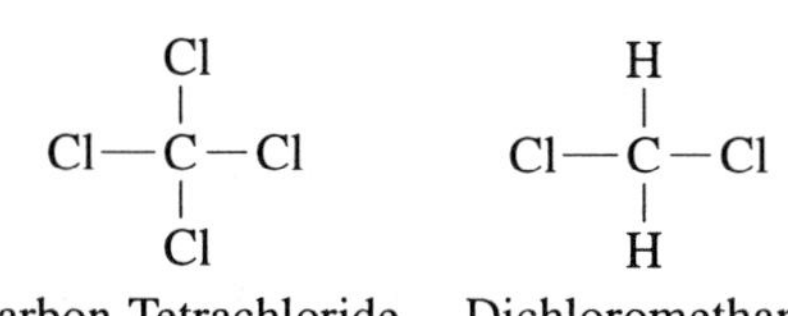

Carbon Tetrachloride Dichloromethane

Alkenyl halides contain at least one halogen atom and at least one carbon–carbon double bond. Examples include trichloroethylene (TCE), C_2HCl_3, and tetrachloroethylene or perchloroethylene (perc), C_2Cl_4.

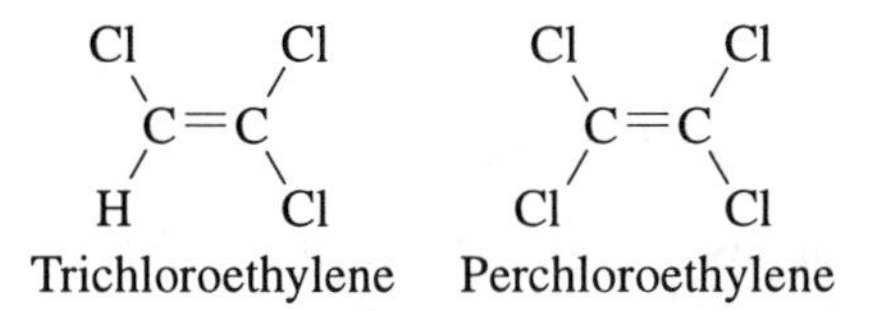

Trichloroethylene Perchloroethylene

Table 2.2 gives the physical properties of several chlorinated hydrocarbons that are widely used and often found as contaminants in groundwater (U.S. EPA, 1990b).

Table 2.2 Physical Properties for Selected Chlorinated Hydrocarbons

Aromatic Hydrocarbons	Water Solubility (mg/L)	Vapor Pressure (mmHg)	Henry's Constant (atm-m^3/mol)
Chloroethane (Ethyl Chloride)	1.75E+03	9.52E+01	5.59E–03
Chloroethene (Vinyl Chloride)	5.70E–03	2.20E–08	1.16E–06
Chloromethane (Methyl Chloride)	1.20E–03	5.60E–09	1.55E–06
Dibromomethane	4.46E+02	4.14E+00	1.92E–03
Dichlorodifluoromethane (Freon 12)	4.66E+02	1.17E+01	3.72E–03
1,1-Dichloroethane	3.30E+03	1.00E+00	5.06E–05
1,2-Dichloroethane	1.00E+02	1.00E+00	1.93E–03
1,1-Dichloroethene	1.23E+02	2.28E+00	3.59E–03
cis-1,2-Dichloroethene	7.90E+01	1.18E+00	2.89E–03
trans-1,2-Dichloroethene	4.60E+03	5.90E–02	2.75E–06
Dichloromethane (Methylene Chloride)	2.50E+00	3.00E–01	2.54E–02
Pentachloroethane	4.20E+03	6.21E–02	2.38E–06
1,1,1,2-Tetrachloroethane	5.60E+03	1.49E–05	6.45E-10
1,1,2,2-Tetrachloroethane	2.40E+02	5.10E–03	5.09E–06

Table 2.2 Physical Properties for Selected Chlorinated Hydrocarbons (Continued)

Aromatic Hydrocarbons	Water Solubility (mg/L)	Vapor Pressure (mmHg)	Henry's Constant (atm-m³/mol)
Tetrachloroethane (PERC)	1.32E+03	1.80E–02	3.27E–06
Tetrachloromethane	1.52E+02	7.00E+00	6.43E–03
Tribromomethane (Bromoform)	6.00E–03	1.09E–05	6.81E–04
1,1,1-Trichloroethane	3.17E+01	2.30E–01	1.15E–03
1,1,2-Trichloroethane	1.90E+03	1.50E–01	2.20E–05
Trichloroethene (TCE)	7.11E–02	1.13E–04	6.18E–04
Trichlorofluoromethane (Freon 11)	1.40E+01	1.10E–04	2.75E–06
Trichloromethane (Chloroform)	9.30E+04	3.41E–01	4.54E–07

2.2.5 Nitrogen- and Sulfur-Containing Hydrocarbons

Some common industrial chemicals that fall under the nitrogen-containing hydrocarbon group include 2- and 4-nitrotoluene, 2,3-dinitrotuluene, and 2,4,6-trinitrotoluene (TNT). TNT is an explosive and is a common soil and groundwater contaminant in ammunitions and explosives waste disposal areas. The other famous nitrogen-containing hydrocarbon is Atrazine, a complex molecule that is widely used as an herbicide for weed control.

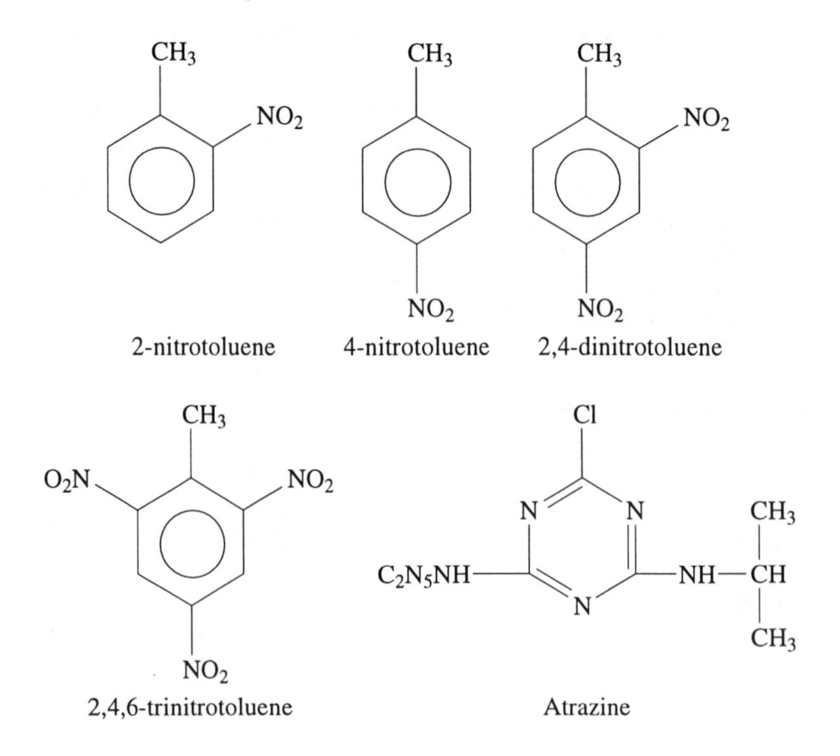

Most common sulfur-containing hydrocarbons are mercaptans. Typically, sulfur-containing hydrocarbons have offensive odors. Many sulfur-containing hydrocarbons are utilized as pesticides. For example, captan, malathion, and asulam, which are used as fungicide, insecticide, and herbicide, respectively.

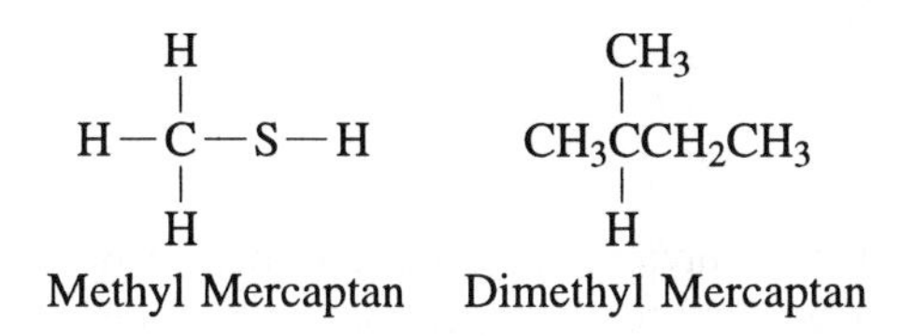

Methyl Mercaptan Dimethyl Mercaptan

2.2.6 Degradation of Hydrocarbons

As with many things that exist in this world, hydrocarbons degrade. They can either undergo microbial degradation or degradation by chemical reactions to form methane or carbon dioxide and water. Hydrocarbons can be degraded under both aerobic and anaerobic conditions by microbes. Some bacteria that do this include *Micrococcus, Pseudomonas, Mycobacterium,* and *Nocardia.*

Hydrocarbons can also be degraded by chemical reactions. Some common chemical reactions involved in hydrocarbon degradation are substitution, dehydrogenation, oxidation, and reduction.

2.2.6.1 Substitution

Substitution is also called hydrolysis. It is a reaction where the halogenated functional groups react with water to create an alcohol. During the reaction, the halogenated compound is replaced by a hydroxyl (OH^-) group. The following is an example of substitution reaction:

$$\underset{\text{1-chlorobutane}}{CH_3CH_2CH_2CH_2Cl} + H2O \rightarrow \underset{\text{Butanol}}{CH_3CH_2CH_2CH_2OH} + HCl$$

The reaction rate of the substitution reaction is inversely proportional to the number of halogen ions present. The reaction rate decreases rapidly with increasing number of halogen ions.

2.2.6.2 Dehydrogenation

Dehydrogenation is a reaction in which a halide ion and a hydrogen atom from the adjacent carbon are lost, resulting in a formation of a double bond between the carbon atoms. Dehalogenation can transform 1,1,1-trichloroethane to 1,1-dichloroethene:

$$\underset{\text{1,1,1 - trichloroethane}}{CCl_3CH_3} \rightarrow \underset{\text{1,1 - dichloroethene}}{CH_2 = Cl_2} + HCl$$

The reaction rate of the dehydrogenation reaction is proportional to the number of halogen ions present. The reaction rate increases with increasing number of halogen ions.

2.2.6.3 Oxidation

Oxidation reactions involve the replacement of a hydrogen atom on a carbon that is also attached to a halogen ion, with an OH^-. The result is the formation of a chlorinated alcohol. The oxidation of 1,1-dichloroethane forms 1,1-dichloroethanol:

$$\underset{\text{1,1,1 - dichloroethane}}{CH_3CHCl_2} + H_2O \rightarrow \underset{\text{1,1 - dichloroethanol}}{CH_3CCl_2OH} + 2H^+ + 2e^-$$

2.2.6.4 Reduction

Reduction is a two-stage reaction. It starts with the removal of a halogen ion by a radical species, such as a transition metal complex. The resulting alkyl radical can then react with an H^+ ion, which takes the place of the departed halogen ion. The reduction of 1,1,1-trichloroethane forms 1,1-dichloroethane:

$$\underset{\text{1,1,1 - trichloroethane}}{CH_3CCl_3} + H^+ + e^- \rightarrow \underset{\text{1,1 - dichloroethane}}{CH_3CHCl_2} + Cl^-$$

Reduction can also occur if there are two halogen ions on adjacent carbon atoms. In this case, the loss of a halogen from each carbon creates a double bond between the carbon atoms. The reduction of hexachloroethane forms tetrachloroethene:

$$\underset{\text{hexachloroethane}}{CCl_3CCl_3} + H^+ + e^- \rightarrow \underset{\text{tetrachloroethane}}{CH_3CCl_2OH} + Cl^-$$

2.3 CLASSES OF CONTAMINANTS

Generally, the groups of contaminants that a remediation engineer will typically encounter can be classified as petroleum products and halogenated hydrocarbons. Both petroleum products and halogenated hydrocarbons can be soluble in water and thus found as contaminants in groundwater.

If spilled on the ground in quantities great enough to overcome residual saturation, the pure phase petroleum products will float on the water table and are referred to as Light Non-Aqueous Phase Liquids (LNAPL). On the other hand, the halogenated hydrocarbons may migrate vertically downward through an aquifer because they are denser than water. They are referred to as "sinkers" or Dense Non-Aqueous Phase Liquids (DNAPL).

DNAPL contamination is more difficult to remediate than LNAPL contamination simply because of their densities relative to the water density. More complications result when dealing with DNAPL remediation because DNAPL are (i) not nearly as biodegradable as other organic compounds, (ii) nonsorbing and therefore quite mobile in a groundwater system, and (iii) rather volatile.

With respect to their densities, when spilled in adequate volumes, these compounds are capable of penetrating the capillary fringe and sinking deep into an aquifer system. In addition, the low viscosities of many of these compounds serve to facilitate both their initial percolation through the unsaturated zone and their subsequent entry into the saturated zone.

Solvents spilled on the ground surface are thought to be capable of volatilizing rapidly and completely to the atmosphere. Such volatilization was indeed important in reducing the severity of many cases of solvent contamination. However, it must be realized that there is potential for volatilization and diffusion into the unsaturated zone because of the driving force created by the absence of these chemicals in the unsaturated zone. In addition, the presence of moisture in the unsaturated zone could facilitate the spreading and increase the magnitude of subsurface contamination.

2.4 HYDROCARBON PRODUCTS FROM PETROLEUM

Petroleum is composed chiefly of saturated (paraffin) hydrocarbons, but may also contain unsaturated hydrocarbons, aromatic hydrocarbons and their derivatives, nitrogen compounds, and sulfur compounds. The main step in the refining of petroleum is separation by distillation into a number of fractions, each of which is a complex mixture of hydrocarbons and has properties that make it commercially valuable.

The composition of each fraction depends upon the temperature range over which it is collected. Some of the more important products obtained from refining petroleum, classified according to number of carbon atoms, are listed in Table 2.3. Before the distillation products of petroleum are ready for use, further purification is usually necessary.

Regular gasoline is a mixture of more than 100 organic compounds. Table 2.4 gives an analysis of fresh and weathered gasolines (Johnson et al., 1990). In addition to the listed chemicals, there are many compounds present that constitute less than 0.01% of the mixture.

Gasolines are rated on an arbitrary scale in which iso-octane, once thought to be the ideal fuel for the gasoline engine, is given the rating "100 octane."

Table 2.3 Hydrocarbon Products from Petroleum

	Approximate Number of Carbon Atoms	Boiling Point Range (C)	Uses
Petroleum Ether	4 to 10	35–80	Solvent
Gasoline	4 to 13	40–225	Fuel
Kerosene	10 to 16	175–300	Fuel, Lighting
Lubricating Oils	< 20	< 350	Lubrication
Paraffin	23 to 29	50–60 (m.p.)	Candles, Wax
Asphalt		Viscous Liquids	Paving, Roofing
Coke		Solid	Solid Fuel

Table 2.4 Composition (Mass Fraction) of Fresh and Weathered Gasolines

Compound Name	Molecular Weight (g)	Fresh Gasoline	Weathered Gasoline	Approximate Composition
propane	44.1	0.0001	0.0000	0
isobutane	58.1	0.0122	0.0000	0
n-butane	58.1	0.0629	0.0000	0
trans-2-butene	56.1	0.0007	0.0000	0
cis-2-butene	56.1	0.0000	0.0000	0
3-methyl-1-butene	70.1	0.0006	0.0000	0
isopentane	72.2	0.1049	0.0069	0.0177
1-pentene	70.1	0.0000	0.0005	0
2-methyl-1-butene	70.1	0.0000	0.0008	0
2-methyl-1,3-butadiene	68.1	0.0000	0.0000	0
n-pentane	72.2	0.0586	0.0095	0
trans-2-pentene	70.1	0.0000	0.0017	0
2-methyl-2-butene	70.1	0.0044	0.0021	0
2-methyl-1,2-butadiene	68.1	0.0000	0.0010	0
3,3-dimethyl-1-butene	84.2	0.0049	0.0000	0
cyclopentane	70.1	0.0000	0.0046	0.0738
3-methyl-1-pentene	84.2	0.0000	0.0000	0
2,3-dimethylbutane	86.2	0.0730	0.0044	0
2-methylpentane	86.2	0.0273	0.0207	0
3-methylpentane	86.2	0.0000	0.0186	0
n-hexane	86.2	0.0283	0.0207	0
methylcyclopentane	84.2	0.0083	0.0234	0
2,2-dimethylpentane	100.2	0.0076	0.0064	0
benzene	78.1	0.0076	0.0021	0
cyclohexane	84.2	0.0000	0.0137	0.1761
2,3-dimethylpentane	100.2	0.0390	0.0000	0
3-methylhexane	100.2	0.0000	0.0355	0

Table 2.4 Composition (Mass Fraction) of Fresh and Weathered Gasolines (Continued)

Compound Name	Molecular Weight (g)	Fresh Gasoline	Weathered Gasoline	Approximate Composition
3-ethylpentane	100.2	0.0000	0.0000	0
n-heptane	100.2	0.0063	0.0447	0
2,2,4-trimethylpentane	114.2	0.0121	0.0503	0
methylcyclohexane	98.2	0.0000	0.0393	0
2,2-dimethylhexane	114.2	0.0055	0.0207	0
toluene	92.1	0.0550	0.0359	0.1926
2,3,4-trimethylpentane	114.2	0.0121	0.0000	0
3-methylheptane	114.2	0.0000	0.0343	0
2-methylheptane	114.2	0.0155	0.0324	0
n-octane	114.2	0.0013	0.3000	0
2,4,4-trimethylhexane	128.3	0.0087	0.0034	0
2,2-dimethylheptane	128.3	0.0000	0.0226	0
ethylbenzene	106.2	0.0000	0.0130	0
p-xylene	106.2	0.0957	0.0151	0
m-xylene	106.2	0.0000	0.0376	0.1641
3,3,4-trimethylhexane	128.3	0.0281	0.0056	0
o-xylene	106.2	0.0000	0.0274	0
2,2,4-trimethylheptane	142.3	0.0105	0.0012	0
n-nonane	128.3	0.0000	0.0382	0
3,3,5-trimethylheptane	142.3	0.0000	0.0000	0
n-propylbenzene	120.2	0.0841	0.0117	0.1455
2,3,4-trimethylheptane	142.3	0.0000	0.0000	0
1,3,5-trimethylbenzene	120.2	0.0411	0.0493	0
1,2,4-trimethylbenzene	120.2	0.0213	0.0707	0
n-decane	142.3	0.0000	0.0140	0
methylpropylbenzene	134.2	0.0351	0.0170	0
dimethylethylbenzene	134.2	0.0307	0.0289	0.0534
n-undecane	156.3	0.0000	0.0075	0
1,2,4,5-tetramethylbenzene	134.3	0.0133	0.0056	0
1,2,3,4-tetramethylbenzene	134.2	0.0129	0.0704	0.1411
1,2,4-trimethyl-5-ethylbenzene	148.2	0.0405	0.0651	0
n-dodecane	170.3	0.0230	0.0000	0
naphthalene	128.2	0.0045	0.0076	0
n-hexylbenzene	162.3	0.0000	0.0147	0.0357
methylnaphthalene	142.2	0.0023	0.0134	0
Total		1.0000	1.0000	1.0000

Gasolines with octane numbers less than 100 are less efficient than iso-octane and those with octane numbers greater than 100 are more efficient.

Gasoline has usually been made by blending aliphatic and aromatic hydrocarbons and adding tetraethyl lead. However, manufacturers now are producing gasoline without adding tetraethyl lead because of possible toxic effects of lead compounds in the atmosphere and the poisoning effect of tetraethyl lead on the catalysts in catalytic converters.

Currently, gasoline is being produced without tetraethyl lead by blending in more highly branched isomers. In addition, oxygenated fuel, which burns more efficiently and produces less air pollutants, is being produced by blending ethanol with gasoline.

2.5 LABORATORY ANALYSES

There are numerous methods that are available to analyze organic compounds trapped in soils and dissolved in groundwater. Design engineers should familiarize themselves with the various types of laboratory analyses available. Almost all analyses for hazardous waste are based on methods developed by the EPA. Test methods based on the modification of certain EPA techniques, such as "Modified EPA Method 8015," are also popular in some states.

It should be noted that the EPA methods were not specifically developed for the purpose of analyzing petroleum-contaminated soil and groundwater (Calabrese et al., 1993). Several different techniques exist for the analyses of similar parameters, e.g., Methods 502, 524, 624, 8020, 8240, and 8260 for the analyses of benzene, toluene, ethylbenzene, and xylene (BTEX). Figure 2.1 shows the total petroleum hydrocarbon ranges and the associated analysis methods. In designing remediation systems, the engineer should be familiar with the various methods available.

Before analyses, the samples have to be prepared. The most widely used extraction or concentration process is the purge and trap method, or EPA Method 5030. For the direct measurement of specific contaminants in petroleum-contaminated soil and groundwater, the EPA 500, 600, and 8000 series are the most widely used methods. The 500 and 600 methods are used for the analyses of water samples. The 8000 series is applicable for both soil and water samples. Table 2.5 presents some of the various methods.

The methods listed in Table 2.5 test for specific constituents. Methods for the analyses of indicator parameters also see wide application. EPA Method 418.1, Total Recoverable Petroleum Hydrocarbons (TRPH), is used in the determination of the presence of petroleum compounds in both soil and water samples. EPA Method 413.1, Oil and Grease, sees more application in the wastewater industry. Modified Method 8015 is a little more specific than 418.1 in that it differentiates between lighter gasoline products and heavier diesel fuels.

Total volatile petroleum hydrocarbons (TVPH) or total petroleum hydrocarbons as gasoline (TPHg) is used as a measure of the total concentration

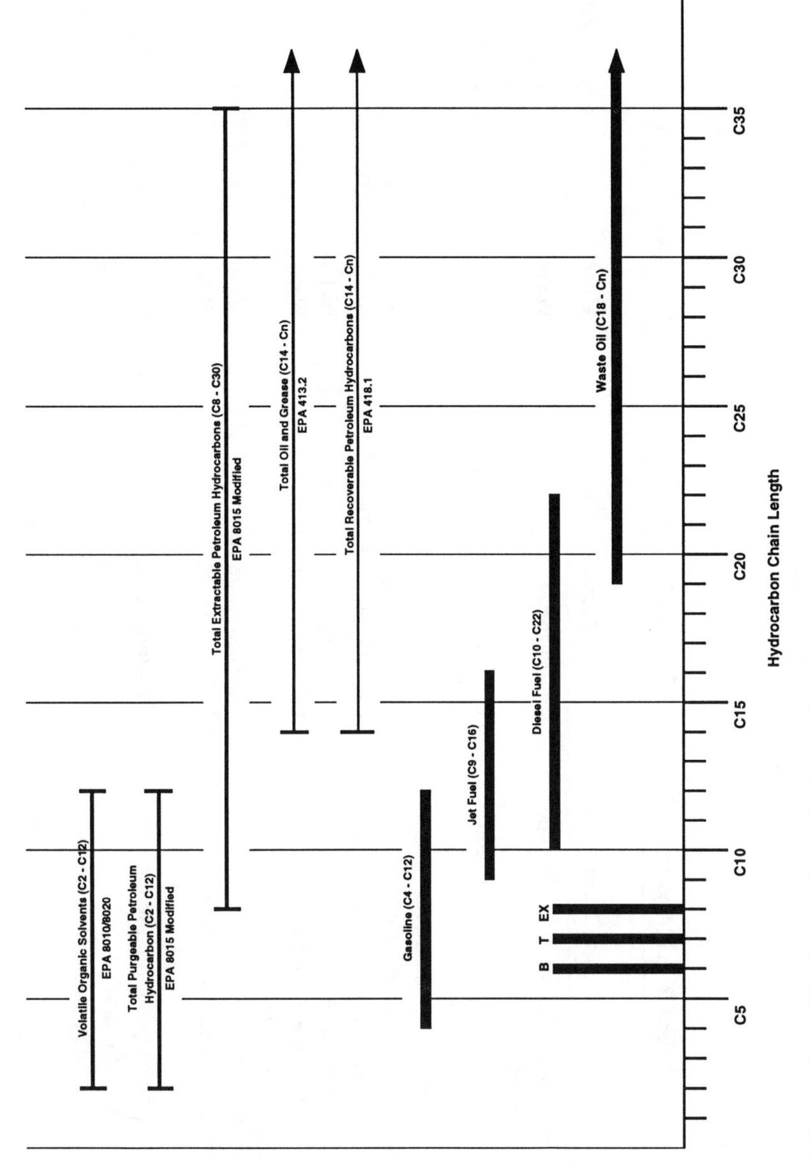

Figure 2.1. Total petroleum hydrocarbon ranges.

Table 2.5 Analytical Methods for Petroleum-Contaminated Soils and Groundwater

Method No.	Method Name	Method of Analysis	Matrix	Parameters
502.1	Volatile Halogenated Organic Compounds	GC-Purge & Trap - Hall Detector	Water	1,2-, 1,3-, 1,4-Dichlorobenzene, CCl_4, etc.
502.2	Volatile Organic Compounds	GC-Purge & Trap - Hall Detector	Water	BTEX, CCl_4, etc.
503.1	Volatile Aromatic & Unsaturated Organic Compounds	GC-Purge & Trap - PID	Water	BTEX, etc.
524.1	Volatile Organic Compounds	GC/MS-Purge & Trap	Water	BTEX, CCl_4, etc.
524.2	Volatile Organic Compounds	GC/MS-Purge & Trap	Water	BTEX, etc.
602	Purgeable Aromatics	GC-Purge & Trap - PID	Water	BTEX, etc.
610	Polynuclear Aromatic Hydrocarbons	GC-Extraction - FID	Water	
624	Purgeables	GC/MS-Purge & Trap - PID	Water	BTEX, CCl_4, TCE, etc.
625	Base/Neutrals & Acids	GC/MS-Extraction	Water	1,3- & 1,4-Dichlorobenzene, PCB, etc.
8010	Halogenated Volatile Organic Compounds	GC-Purge & Trap - Hall Detector	Soil Water	CCl_4, TCE, PCE, etc.
8015	Nonhalogenated Volatile Organic Compounds	GC-Purge & Trap - FID	Soil Water	MEk, MIBK, Ethanol, etc.

8020	Aromatic Volatile Hydrocarbons	GC-Purge & Trap - PID	Soil Water	BTEX, Chlorobenzene, 1,2-, 1,3-, 1,4-Dichlorobenzene
8100	Polynuclear Aromatic Hydrocarbons	GC-Direct Injection - FID	Soil Water	Naphthalene, Pyrene, etc.
8120	Chlorinated Hydrocarbons	GC-Direct Injection - ECD Detector	Soil Water	1,2-, 1,3-, & 1,4-Dichlorobenzene, etc.
8240	Volatile Organic Compounds	GC/MS-Purge & Trap	Soil Water	BTEX, CCl_4, etc.
8260A	Volatile Organic Compounds	MS-Extraction	Soil Water	BTEX, CCl_4, etc.
8270	Semi-Volatile Organic Compounds	GC/MS-Extraction	Soil Water	Phenol, Naphthalene, etc.
8310	Polynuclear Aromatic Hydrocarbons	HPLC-Extraction-UV & Fluorescence Detectors	Soil Water	Naphthalene, Pyrene, etc.
418.1	Total Recoverable Petroleum Hydrocarbon	IR	Soil Water	Petroleum Hydrocarbons
8015M	Total Petroleum Hydrocarbon as Gasoline	GC	Soil Water	Gasoline Compounds
8015M	Total Petroleum Hydrocarbon as Diesel	GC	Soil Water	Diesel Compounds

of gasoline compounds in the soil or groundwater. Likewise, total extractable petroleum hydrocarbons (TEPH) or total petroleum hydrocarbons as diesel (TPHd) is used as an indicator of the total concentration of diesel products in the soil or groundwater. Samples for TPHg and TPHd analyses are prepared by EPA Method 5030 (Purge and Trap) and extracted by sonification using EPA Method 3550, respectively (Tri-Regional Board Staff Recommendations, 1990).

The analysis of soil and water samples will often show that a particular compound was not found. The correct number, however, is not reported as zero. Instead, it is reported as non detect, with the detection limit given in parentheses next to it. The detection limit is the lowest concentration level that can be determined to be statistically different than a blank sample (American Chemical Society Committee on Environmental Improvement, 1983). A method detection limit is specified for each of the EPA methods.

Quality assurance and quality control (QA/QC) procedures will need to be implemented in the analysis of soil and water for contamination to ensure the accuracy (correctness) and the precision (reproducibility) of the data. The typical QA/QC procedure will involve the collection of field blanks and field duplicate samples during the sampling event and analyses of method blanks and spiked samples in the laboratory.

Field blanks are collected during the sampling event by running clean distilled water through the sampling equipment and collecting the water in a bottle for lab analysis. The purpose of field blanks is to determine if any organics are being introduced during the sampling and analysis procedures. A field blank should contain no organic compounds.

Field duplicate samples are used to check the precision of the analysis. Usually, two or three identical samples are collected from a sampling point and submitted to the lab as blind duplicates so that the lab personnel do not know that it is a duplicate sample. Sometimes, the lab takes a field sample, splits it into two aliquots, and analyzes them individually to check for the precision of the analysis.

Method blanks are actually distilled water analyzed for the solvents and chemicals called for in the analytical procedure. This is to test the purity of the solvents and reagents used in the sample analysis.

Spiked samples are samples containing a known amount of organic compound that was added to the water sample collected in the field. Usually, noncontaminated groundwater from the site is collected and spiked with a known amount of organic compound and analyzed. The percent recovery of the spiked organic compound can be determined with this procedure because the initial concentration of the organic compound is known. A perfect analysis will have a 100% recovery. Typically, the acceptable percent recovery falls between 95 and 105%.

3 BASIC GEOLOGY AND HYDROGEOLOGY

3.1 INTRODUCTION

This chapter reviews the fundamental concepts and principles of geology and hydrogeology. A basic working knowledge of the various types of soil and their properties and the hydrogeological characteristics of aquifers as well as the transport mechanisms of contaminants in the subsurface is important to design engineers because they are dealing with contamination that occurs beneath the ground. The type of remediation system the engineer selects and designs is ultimately dictated by the subsurface characteristics of the site. The fate and transport of contaminants in the subsurface are also discussed in this chapter.

3.2 REVIEW OF GEOLOGY AND SOIL TYPES

3.2.1 Soil Formation and the Nature of Soil Constituents

The earth has a crust of granitic and basaltic rock 40 to 130 ft thick. Overlying this solid rock is a relatively thin layer of variable thickness of unconsolidated materials called soil. These materials can vary in size from submicroscopic mineral particles to huge boulders. These unconsolidated materials were formed as a result of weathering and other geologic processes acting on the rocks near the earth's surface. There are basically two types of weathering processes, physical weathering and chemical weathering.

Physical weathering, also called mechanical weathering, causes disintegration of rocks into smaller-sized particles. Physical weathering agents include freezing and thawing, temperature changes, erosion, and the activity of plants and animals. Chemical weathering decomposes the minerals that make up rocks through reactions such as oxidation, reduction, carbonation, and other chemical processes. Soils present at a particular site can be either residual (weathered in place) or transported (moved by natural elements).

Table 3.1 Soils Characteristics

Soil Name	Grain Size	Characteristics
Gravels, Sands	Coarse grained; can see individual grain by eye	Noncohesive; nonplastic; granular
Silts	Fine grained; cannot see individual grains	Some cohesive; some cohesionless
Clays	Fine grained; cannot see individual grains	Cohesive; plastic

3.2.2 Soil Texture

Soil texture is the appearance of a soil. It depends on the relative sizes and shapes of the soil particles as well as the range or distribution of those sizes. Texturally, soils may be divided into coarse-grained or fine-grained soils. A convenient dividing line is the smallest grain that is visible to the naked eye. Soils with particles larger than this size (approximately 0.05 mm) are considered coarse-grained, while soils finer than this size are called fine-grained. Gravels and sands are examples of coarse-grained soils while silts and clays are examples of fine-grained soils.

Another convenient way to classify soils is according to their plasticity and cohesion. For example, sands are nonplastic and noncohesive, whereas clays are both plastic and cohesive. Silts fall between clays and sands; they are fine-grained and yet nonplastic and noncohesive. These relationships and some general soil characteristics are presented in Table 3.1.

3.2.3 Grain Size and Grain Size Distribution

The range of possible particle sizes in soils is tremendous. Soils can range from boulders several centimeters in diameter down to ultrafine-grained colloidal materials. Because the maximum possible range is on the order of 10^8, grain size distributions are typically plotted against the logarithm of the average grain diameter. Figure 3.1 indicates the divisions between the various textural sizes according to several common classification schemes (Robert et al., 1981).

The grain size is obtained by conducting a gradation test. For coarse-grained soils, a sieve analysis is performed in which a sample of dry soil is shaken mechanically through a series of woven-wire square-mesh sieves with successively smaller openings. Since the total mass of sample is known, the percentage retained or passing each size sieve can be determined by weighing the amount of soil retained on each sieve after shaking. The U.S. standards sieve numbers for the particle size analysis of soils are shown in Table 3.2 (Robert et al., 1981).

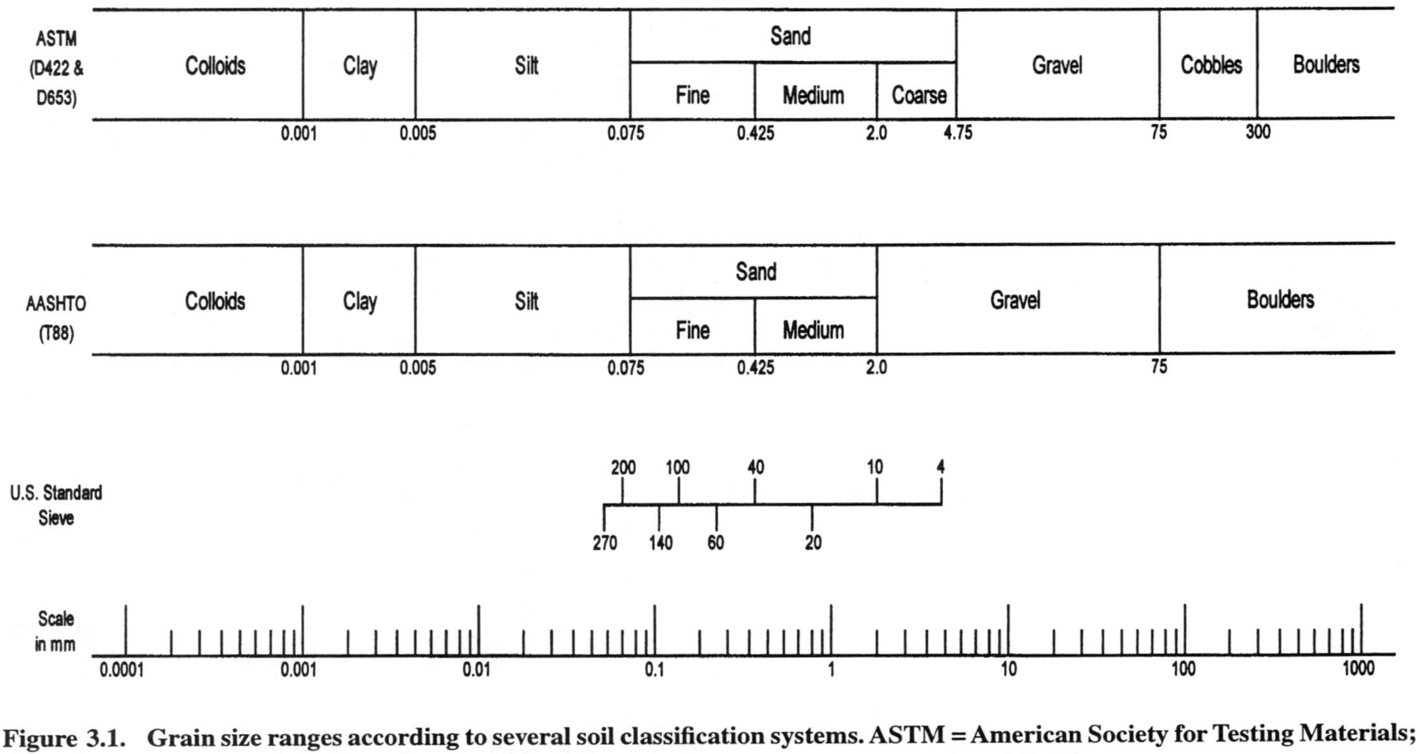

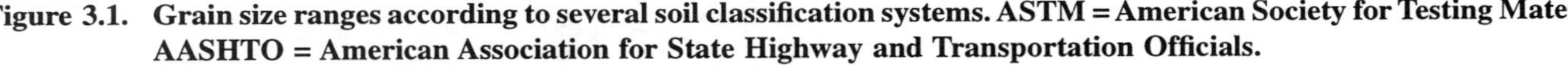

Figure 3.1. Grain size ranges according to several soil classification systems. ASTM = American Society for Testing Materials; AASHTO = American Association for State Highway and Transportation Officials.

Table 3.2 U.S. Standard Sieve Sizes and Their Corresponding Open Dimension

U.S. Standards Sieve No.	Sieve Opening (mm)
4	4.75
6	3.33
8	2.36
10	2.00
20	0.833
30	0.589
40	0.425
50	0.295
60	0.250
70	0.208
80	0.177
100	0.149
140	0.106
200	0.075
700	0.038

For fine-grained soils, the hydrometer analysis is used. The hydrometer analysis is based on Stoke's law for falling spheres in a viscous fluid in which the terminal velocity of fall depends on the grain diameter and the density of the grains in suspension and of the fluid. The grain diameter can then be calculated from a knowledge of the distance and time of fall. The hydrometer also determines the specific gravity of the suspension, and this enables the percentage of particles of a certain equivalent particle diameter to be calculated.

The distribution of the percentage of the total sample less than a certain sieve size is typically plotted in a cumulative frequency diagram. The equivalent grain sizes are plotted to a logarithmic scale on the abscissa, whereas percentage by weight of the total sample either passing or retained is plotted linearly on the ordinate. Some typical grain size distributions are shown in Figure 3.2. Well-graded soil has a good representation of particle sizes over a wide range, and its gradation curve is smooth and generally concaves upwards. On the other hand, poorly graded soil will have an excess or deficiency of certain sizes, or most of the particles will be about the same size. The uniform and gap-graded soils in Figure 3.2 are examples of poorly graded soils.

The letter D is used to represent equivalent "percent passing" on the grain size distribution curve. For example, D_{10} is the grain size that corresponds to 10% of the sample passing by weight, which means 10% of the particles are smaller than the diameter D_{10}. This parameter is called the effective size.

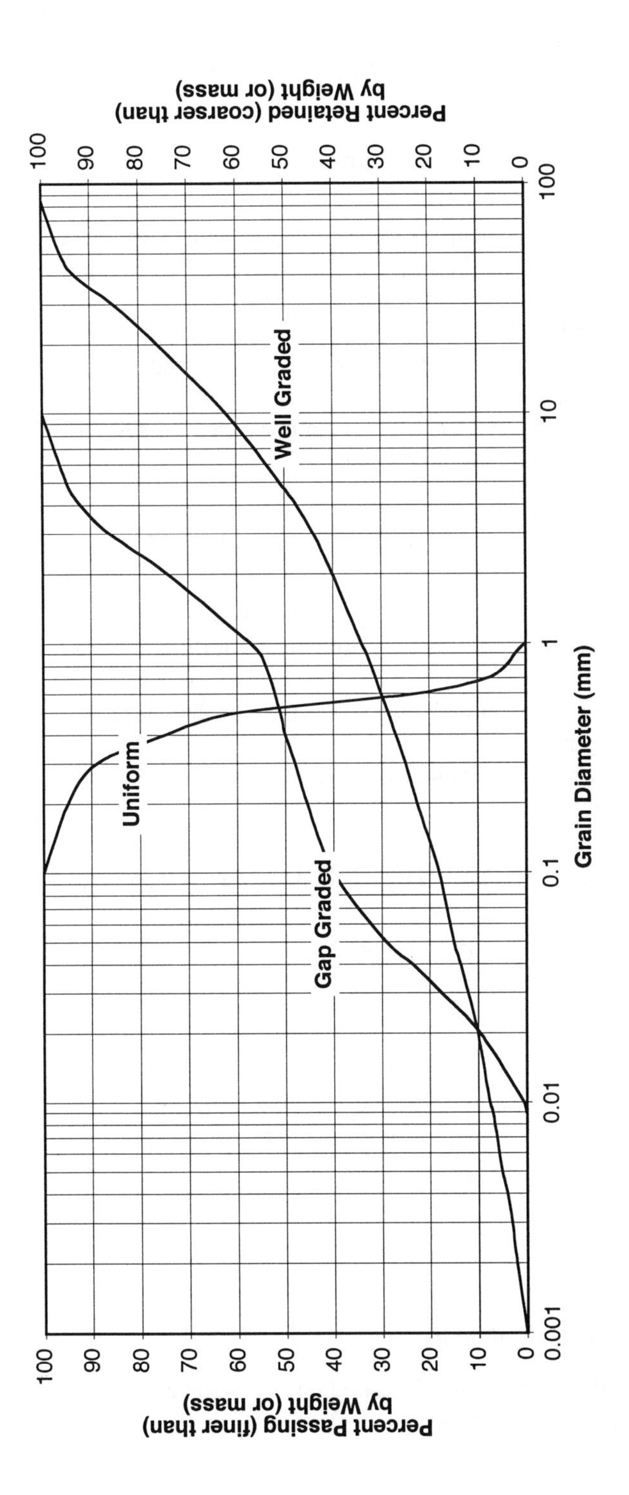

Figure 3.2. Typical grain size distributions.

The coefficient of uniformity C_u is a crude shape parameter, and it is defined as

$$C_u = D_{60}/D_{10}$$

where D_{60} is the grain diameter (in mm) corresponding to 60% passing and D_{10} is the grain diameter (in mm) corresponding to 10% passing.

The smaller the C_u, the more uniform the gradation. Very poorly graded soil will have a relatively small C_u while very well-graded soils will have a relatively large C_u.

Another shape parameter that is sometimes used for soil classification is the coefficient of curvature. It is defined as

$$C_c = (D_{30})^2/(D_{10})(D_{60})$$

A soil with a coefficient of curvature between 1 and 3 is considered to be well graded as long as C_u is also larger than 4 for gravels and 6 for sands.

Example: Determine D_{10}, C_u, and C_c for the distribution of uniform soil in Figure 3.2.
Pick off the diameters corresponding to 10%, 30%, and 60% passing.

$$D_{10} = 0.3 \text{ mm} \qquad D_{30} = 0.43 \text{ mm} \qquad D_{60} = 0.55 \text{ mm}$$

$$\begin{aligned} C_u &= D_{60}/D_{10} \\ &= 0.55/0.3 \\ &= 1.8 \end{aligned}$$

$$\begin{aligned} C_c &= (D_{30})^2/(D_{10})(D_{60}) \\ &= (0.43)^2/(0.3)(0.55) \\ &= 1.12 \end{aligned}$$

3.2.4 Classification of Soils

Soil classification is generally based on soil particle size. Four groups of soils, namely, gravels, sands, silts, and clays make up the basic types of soils.

The three most well-known classification systems developed are the U.S. Department of Agriculture (USDA) system, the American Association of State Highway and Transportation Officials (AASHTO) system, and the Unified Soil Classification System (USCS).

The USCS is one of the two most widely used systems for classifying soils and has been adopted by the American Society for Testing Materials (ASTM). The system is illustrated in Tables 3.3a (ASTM, 1984), Tables 3.3 b–f, and Figure 3.3 (ASTM, 1984). The primary classification method of the system basically subdivides soil groups into two broad categories: coarse-grained soils (gravels and sands) and fine-grained soils (silts and clays). The secondary classification procedure is somewhat of a descriptive nature, based on factors such as the gradation and plasticity of the soil.

The USCS utilizes a two-letter symbol to identify and differentiate between the various combinations of soil groups. The first letter correlates with the primary classification and designates the dominant type of soil:

G indicates gravel
S indicates sand
M indicates silt
C indicates clay
O indicates organic

The second letter of the symbol provides additional information on the soil:

W indicates well graded
P indicates poorly graded
M indicates silt
C indicates clay
L indicates low liquid limit
H indicates high liquid limit

For example, a soil type designated SM is described as a silty sand; an ML designation indicates inorganic silts.

The AASHTO system is the other widely used system for classifying soils. The ASSHTO system divides soils into seven major groups, A-1 through A-7. Groups A-1, A-2, and A-3 are classified as granular materials that have 35% or less material that passes through the No. 200 sieve. If more than 35% of the material passes the No. 200 sieve, the soils are classified as silty or clayey and are designated as groups A-4, A-5, A-6, or A-7 soils. Table 3.4 summarizes the AASHTO classification system.

Table 3.3a ASTM Soil Classification System (USCS)

Soil Classification Chart: Criteria for Assigning Group Symbols and Group Names Using Laboratory Tests[a]				Group Name[b] Symbol	Soil Classification
			Group		
Coarse-Grained Soils More than 50% retained on No. 200 sieve	Gravels More than 50% of coarse fraction retained on No. 4 sieve	Clean Gravels Less than 5% fines[c]	Cu·4 and 1 µCcµ3[e]	GW	Well-graded gravel[f]
			Cu<4 and/or 1>Cc>3[e]	GP	Poorly graded gravel[f]
		Gravels with Fines More than 12% fines[c]	Fines classify as ML or MH	GM	Silty gravel[f,g,h]
			Fines classify as CL or CH	GC	Clayey gravel[f,g,h]
	Sands 50% or more of coarse fraction passes No. 4 sieve	Clean Sands Less than 5% fines[d]	Cu·6 and 1 µCcµ3[e]	SW	Well-graded sand[i]
			Cu<6 and/or 1>Cc>3[e]	SP	Poorly graded sand[i]
		Sands with Fines More than 12% fines[d]	Fines classify as ML or MH	SM	Silty sand[g,h,i]
			Fines classify as CL or CH	SC	Clayey sand[g,h,i]
Fine-Grained Soils 50% or more passes the No. 200 sieve	Silts and Clays Liquid limit less than 50	Inorganic	PI>7 and plots on or above "A" line[j]	CL	Lean clay[k,l,m]
			PI<4 or plots below "A" line[j]	ML	Silt[k,l,m]
		Organic	$\frac{\text{Liquid limit - oven dried}}{\text{Liquid limit - not dried}} < 0.75$	OL	Organic clay[k,l,m] / Organic silt[k,l,m,o]
	Silts and Clays Liquid limit 50 or more	Inorganic	PI plots on or above "A" line	CH	Fat clay[k,l,m]
			PI plots below "A" line	MH	Elastic silt[k,l,m]
		Organic	$\frac{\text{Liquid limit - oven dried}}{\text{Liquid limit - not dried}} < 0.75$	OH	Organic clay[k,l,m,p] / Organic silt[k,l,m,q]

Highly organic soils	Primarily organic matter, dark in color, and organic odor	PT	Peat

[a] Based on the material passing the 3-in. (77-mm) sieve.

[b] If field sample contained cobbles or boulders, or both, add "with cobbles or boulders or both" to group name.

[c] Gravels with 5 to 12% fines require dual symbols: GW-GM well-graded gravel with silt; GW-GC well-graded gravel with clay; GP-GM poorly graded with silt; GP-GC poorly graded gravel with clay.

[d] Sands with 5 to 12% fines require dual symbols: SW-SM well-graded sand with silt; SW-SC well-graded sand with clay; SP-SM poorly graded sand with silt; SP-SC poorly graded sand with clay.

[e] $Cu = D_{60}/D_{10} \quad Cc = \frac{(D_{30})^2}{D_{10} \times D_{60}}$.

[f] If soil contains ≥15% sand, add "with sand" to group name.

[g] If fines classify as CL-ML, use dual symbol GC-GM or SC-SM.

[h] If fines are organic, add "with organic fines" to group name.

[i] If soil contains ≥15% gravel, add "with gravel" to group name.

[j] If Atterberg limits plot in hatched area, soil is a CL-ML, silty clay.

[k] If soil contains 15 to 29% plus No. 200, add "with sand" or "with gravel," whichever is predominant.

[l] If soil contains ≥30% plus No. 200, predominantly sand, add "sandy" to group name.

[m] If soil contains ≥30% plus No. 200, predominantly gravel, add "gravelly" to group name.

[n] PI ≥4 and plots on or above "A" line.

[o] PI <4 or plots below "A" line.

[p] PI plots on or above "A" line.

[q] PI plots below "A" line.

From *Standard Practice for Description and Identification of Soils*, ASTM D2487-84, 1984. With permission.

Table 3.3b Criteria for Describing Moisture Condition of Clay Soil

Description	Criteria
Dry	Absence of moisture, dusty, dry to the touch
Moist	Damp, slightly wet, moisture content below plastic limit
Wet	Moisture content above the plastic limit
Saturated	Very wet; usually soil is below water table

Table 3.3c Criteria for Describing Moisture Condition of Granular Soil

Description	Criteria
Dry	Absence of moisture, dry to the touch
Moist	Damp but no visible free water
Wet	Visible free water
Saturated	Usually soil is below water table

Table 3.3d Criteria for Describing Consistency of Clay Soil

Density	Penetration Resistance, N Blows per 12 in.
Very soft	Less than 2
Soft	2–4
Medium	4–8
Stiff	8–15
Very stiff	15–30
Hard	Greater than 30

Table 3.3e Criteria for Describing Density of Coarse-Grained Soil

Density	Penetration Resistance, N Blows per 12 in.
Loose	Less than 10
Medium	10–30
Dense	30–50
Very dense	Greater than 50

Table 3.3f Criteria for Describing Strength of Rock

Description	Criteria
Very soft	Permits denting by moderate pressure of the fingers
Soft	Resists denting by the fingers, but can be abraded and pierced to a shallow depth by a pencil point

Table 3.3f Criteria for Describing Strength of Rock (Continued)

Description	Criteria
Moderately soft	Resists a pencil point, but can be scratched and cut with a knife blade
Moderately hard	Resistant to abrasion or cutting by a knife blade, but can be easily dented or broken by light blows of a hammer
Hard	Can be deformed or broken by repeated moderate hammer blows
Very hard	Can be broken only by heavy, and in some rocks repeated, hammer blows

3.3 HYDROCARBON PHASES IN SOILS

Petroleum hydrocarbons that are leaked or spilled into the subsurface can be present in several forms. In the vadose zone, hydrocarbons exist in the following four phases, as illustrated in Figure 3.4:

Non-aqueous phase liquid (NAPL)
Adsorbed phase
Vapor phase
Dissolved phase

In the liquid phase or as an NAPL, hydrocarbons are present as free-flowing liquid hydrocarbons in the pore or interstitial spaces between the soil particles. Hydrocarbon contamination in this phase will continue migrating downwards due to the force of gravity. Some lateral movement will also occur because of capillary forces between the liquid and the soil particles. In the adsorbed or sorbed phase (both terms are frequently used interchangeably), the migrating liquid hydrocarbon becomes sorbed to the soil particles and colloids. In the vapor phase, hydrocarbon vapors are present in the pore or interstitial spaces between the soil particles. Finally, some of the hydrocarbons will dissolve in the soil moisture and exist in the soluble state.

3.4 REVIEW OF HYDROGEOLOGY

3.4.1 Classification of Subsurface Water

Water occurring beneath the ground surface is generally divided by the water table into two broad groups: vadose zone water (water content in soil) and groundwater (water beneath the water table). Figure 3.5 depicts a typical subsurface water classification system. The terms saturated zone and groundwater are arbitrary, since the soils directly above the water table (capillary fringe) may be fully saturated and, as such, are an integral part of the

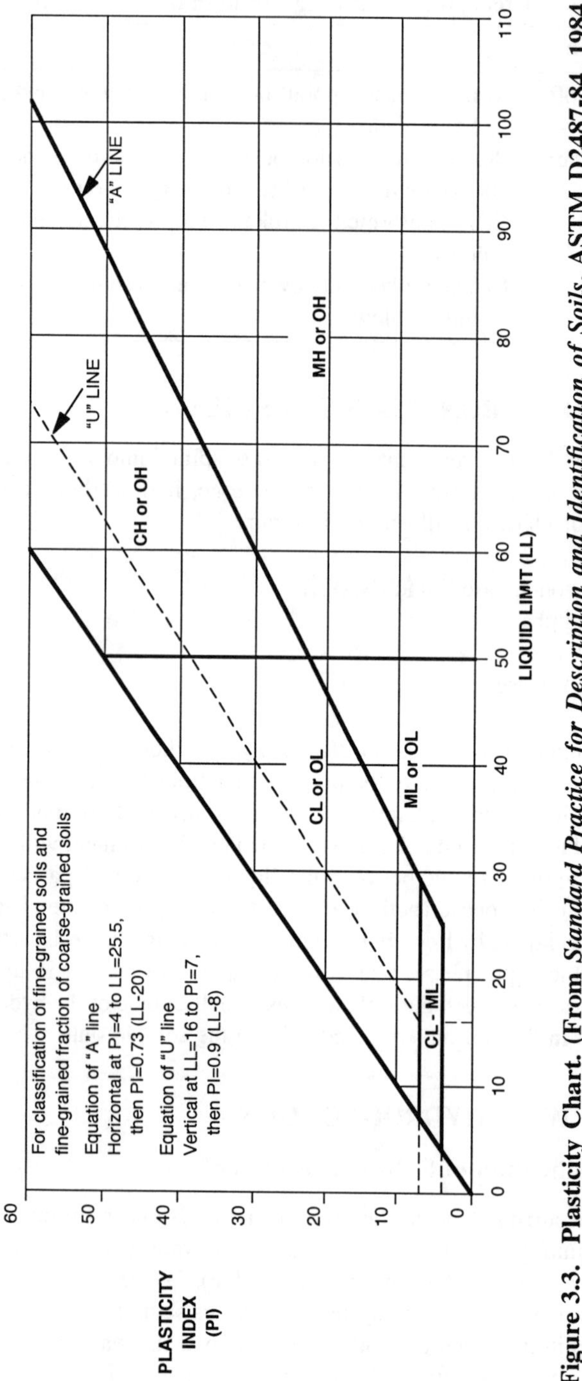

Figure 3.3. Plasticity Chart. (From *Standard Practice for Description and Identification of Soils*, ASTM D2487-84, 1984. With permission.)

Table 3.4 AASHTO Soil Classification System

General classification	Granular Materials (35% or less of total sample passing No. 200)										
	A-1			A-2							
Group classification	**A-1-a**	**A-1-b**	**A-3**	**A-2-4**	**A-2-5**	**A-2-6**	**A-2-7**	**A-4**	**A-5**	**A-6**	**A-7**
Sieve analysis (% passing)											
No. 10	50 max.										
No. 40	30 max.	50 max.	51 min.								
No. 200	15 max.	25 max.	10 max.	35 max.	35 max.	35 max.	35 max.	36 min.	36 min.	36 min.	36 min.
Characteristics of fraction passing No. 40											
Liquid limit				40 max.	41 max.	40 max.	41 max.	40 max.	41 max.	40 max.	41 max.
Plasticity index	6 max.		NP	10 max.	10 max.	11 min.	11 min.	10 max.	10 max.	11 min.	11 min.
Usual types of significant constituent materials	Stone fragments, gravel and sand		Fine Sand	Silty or clayey gravel and sand				Silty soils		Clayey soils	
Generate subgrade rating			Excellent to good						Fair to poor		

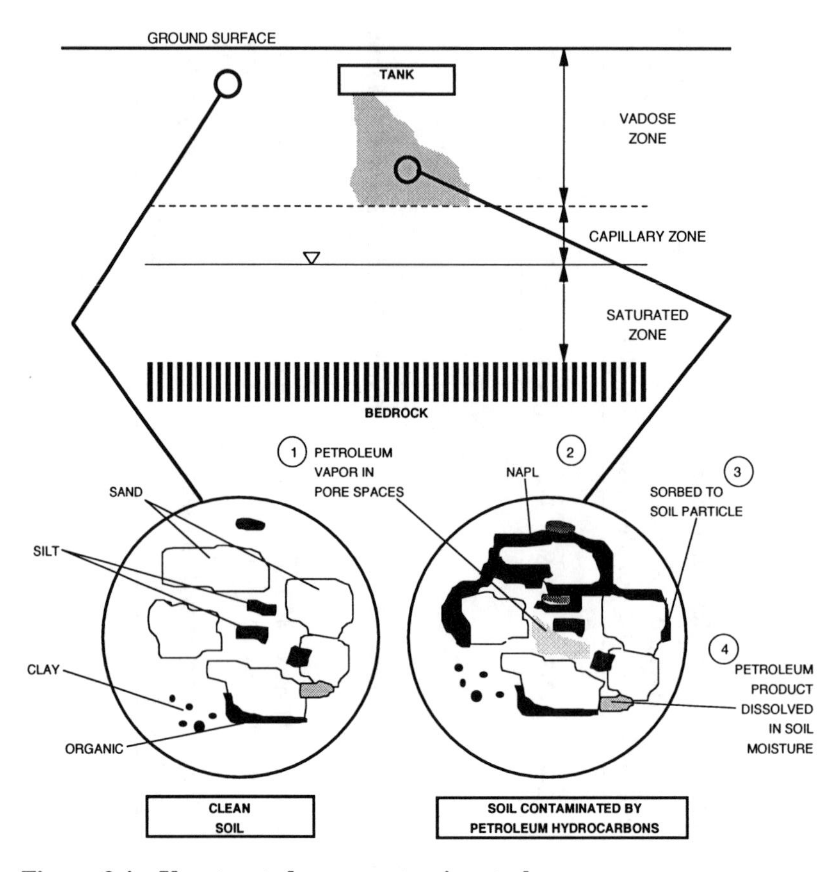

Figure 3.4. Unsaturated zone contaminant phases.

groundwater system. Some authors have accounted for this difference by using the term phreatic zone to describe the zone beneath the water table. In practice, while reviewing hydrogeological reports of contaminated sites, the engineer will seldom encounter the term phreatic zone, but rather, will see the terms saturated zone, water-bearing zone, and aquifer used interchangeably. In this book, no significant distinctions will be made in the usage of these terms.

The distinction between the vadose zone water and the groundwater lies in the difference in fluid pressure. Water above the water table exhibits a fluid pressure that is less than atmospheric pressure. Water beneath the water table has a pressure greater than atmospheric pressure and increases with depth (hydrostatic head of the water above). The water pressure at the water table is equal to atmospheric pressure. If wells are drilled into the saturated or water-bearing zone, the static water level in each well marks the elevation of the groundwater table. Where the water-bearing zone is a confined aquifer, the height of the water column in the well represents the pressure in the aquifer (typically expressed in feet of head).

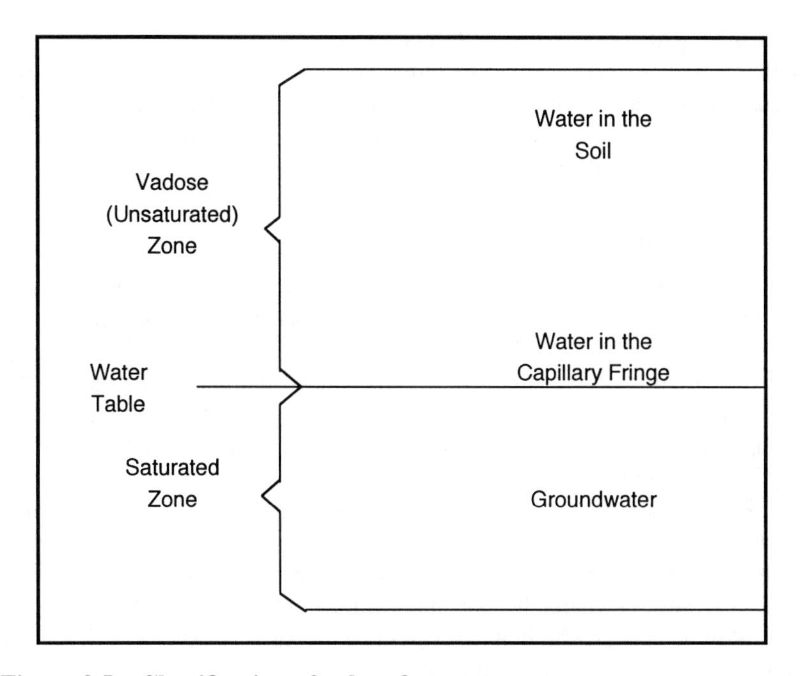

Figure 3.5. Classification of subsurface water.

3.4.2 Aquifers

3.4.2.1 Aquifer

An aquifer, as defined by Freeze and Cherry (1979), is a "saturated permeable unit that can transmit significant quantities of water under ordinary hydraulic gradients." This definition was originally tailored to suit the needs of water-supply hydrogeology. In contaminant hydrogeology, this definition will need to be extended to a broader scope to encompass geological units that are not typically considered to be aquifers in the water supply sense. For example, a predominantly silty clay unit with little groundwater movement might be regarded an aquifer if it were located next to a surface water body that is utilized as a source of drinking water. With such a broad definition, almost any geologic material can be regarded as an aquifer. The only exceptions might be unfractured crystalline rock and massive clay formations.

Materials commonly regarded as aquifers include unconsolidated sedimentary deposits, fractured or porous sandstone, volcanic rocks, and limestone strata. Many other rock types, in certain locations, may also meet the definition of an aquifer.

3.4.2.2 Aquitard

An aquitard is a relatively impermeable body that does not readily transmit water. It typically defines the upper or lower boundary of an aquifer. While an aquifer is defined as a thick formation composed of several stratigraphic units of varying permeability, the term aquitard sometimes refers to the low permeability strata within the aquifer. The term aquifer and aquitard must be viewed as relative; a silt layer might be considered an aquifer in a silty clay unit, and an aquitard in a silty sand unit.

3.4.2.3 Unconfined Aquifer

An unconfined aquifer, also commonly known as a water table aquifer, is an aquifer with a water table as its upper boundary. In an unconfined aquifer, the water table separates the underlying saturated zone from the overlying unsaturated, or vadose, zone. As the unconfined aquifer is defined by the location of the water table, the fluctuation of the water table over time will change the zone of saturation. This change in the zone of saturation plays a significant part in contaminant hydrogeology.

When an unconfined aquifer is underlying contaminated soils, water table fluctuations over time will repeatedly bring the zone of saturation into contact with the contaminated soil, thus further contaminating the groundwater. Each contact serves to recharge the contaminant load in the groundwater that might otherwise be depleted by flow through or, at worst, be supplemented only by contaminants transported via infiltration.

In an unconfined aquifer, draining of some pore spaces accompanies a withdrawal of water from the aquifer. The drainage of pore fluids under groundwater pumping creates a cone of depression in the water table which in turn changes the saturated thickness of the aquifer. For remediation involving free-floating LNAPL, the development of cones of depression leaves behind residual contamination adsorbed onto the soil or otherwise trapped by capillary forces. This residual contamination will serve as a new source of contamination when the pump is shut off and the water table returns to its original position.

The remediation technology commonly employed for this type of contamination is to combine pump and treat technology and soil vapor extraction technology. On the one hand, free-floating product and contaminated groundwater is pumped to the surface for treatment while creating a cone of depression that exposes the contaminated soil that is normally covered by the groundwater. On the other hand, the SVE system reduces the concentration of the contaminants trapped in the soil pores by sweeping fresh air through the soil formation, promoting volatilization of the contaminants.

Most groundwater contamination problems occur initially in unconfined aquifers because of the lack of an overlying low permeability confining layer. This leaves the unconfined aquifer susceptible to direct contamination from liquid chemical spills, lagoons, landfill leachate, and other similar contaminant sources.

3.4.2.4 Confined Aquifer

A confined aquifer is an aquifer that is bound on its upper surface by an aquitard. The upper surface of a confined aquifer remains saturated and the zone of saturation does not fluctuate significantly over time in response to natural or induced changes in the flow system.

In a confined aquifer, two simultaneous processes accompany the withdrawal of groundwater. First, the pore space is reduced due to compaction of the bulk matrix because, as water pressure decreases, stresses on the solid grains increase causing the bulk volume to decrease. Second, the pore water expands due to the reduction in the pressure of the fluid.

3.4.2.5 Perched Aquifer

A special hydrogeological condition is created when a perched aquifer is encountered. A perched aquifer arises when a quantity of water collects on an impermeable stratum above a confined aquifer. Perched aquifers are usually limited in extent and quantity of water, which is recharged by either natural or human sources.

3.4.3 Aquifer Properties

3.4.3.1 Porosity

Porosity is the ratio between the openings (pore spaces) in the soil and the total soil volume as indicated by the formula

$$n = V_v / V_t$$

where n = porosity
V_v = void volume
V_t = total volume

Porosity is usually expressed as a percentage or a decimal fraction. Table 3.5 (Driscoll et al., 1986) depicts typical porosity values associated with various aquifer materials. As indicated in the table, the coarser the material, the lower its porosity. Likewise, the finer-grain materials have greater porosities.

3.4.3.2 Specific Yield

If a mass of saturated soil is allowed to drain by gravity, the ratio of the volume of water drained and the initial total soil mass volume is the specific yield. Specific yield is usually expressed as a percentage

$$S_y = V_d / V_t$$

where S_y = specific yield
V_d = volume of water drained
V_t = total volume

Table 3.5 Porosities for Common Consolidated and Unconsolidated Materials

Unconsolidated Materials	Porosity (%)	Consolidated Materials	Porosity (%)
Clay	45–55	Sandstone	5–30
Silt	35–50	Limestone/Dolomite (original and secondary porosity)	1–20
Sand	25–40	Shale	0–10
Gravel	25–40	Fractured Crystalline Rock	0–10
Sand & Gravel Mixes	10–35	Vesicular Basalt	10–50
Glacial Till	10–25	Dense, Solid Rock	<1

From Driscoll, F. G., *Groundwater and Wells*, Johnson Filtration System Inc., St. Paul, MN, 1986. With permission.

3.4.3.3 Specific Retention

If a mass of saturated soil is allowed to drain by gravity, the ratio of the volume of water retained and the initial total soil mass volume is the specific retention. Specific retention is usually expressed as a percentage

$$S_r = V_r/V_t$$

where S_r = specific yield
V_r = volume of water retained
V_t = total volume

The specific yield plus the specific retention values of a porous medium should equal its porosity

$$n = S_y + S_r$$

Table 3.6 (Heath, 1982) lists the specific yield and specific retention of some selected soil materials.

3.4.3.4 Storage Coefficient

The storage coefficient depends on the compressibility of the water and of the soil structure. It is a dimensionless parameter and is defined as the volume of water that an aquifer releases from or takes into storage per unit surface area of the aquifer per unit change in head

$$S = V/(A \times \Delta h)$$

Table 3.6 Selected Values of Specific Yield and Specific Retention

Material	Specific Yield	Specific Retention
Soil	40	15
Clay	2	48
Sand	22	3
Gravel	19	1
Limestone	18	2
Sandstone (semiconsolidated)	6	5
Granite	0.09	0.01
Basalt (young)	11	3

Note: Values in percent by volume.

where S = storage coefficient
V = volume of water
A = cross-sectional area of the aquifer
Δh = change in head

3.4.3.5 Permeability

The term permeability is sometimes incorrectly used interchangeably with hydraulic conductivity. Although both terms provide an indication as to how easily a particular soil type will allow water to flow through it, i.e., the lower the number, the more difficult or the slower the groundwater flowrate, there are differences in definitions of the two terms. Permeability refers to the capacity of a soil to permit the flow of any fluid through it. It is the function or property of the soil medium only and does not include the effects or properties of the fluid.

Permeability is described by the formula

$$k = K\rho g/u$$

where k = permeability
K = hydraulic conductivity
ρ = density of the fluid
g = acceleration of gravity
u = dynamic viscosity of the fluid

Permeability has the dimensions [L^2] and is expressed in terms of m^2, ft^2, or darcys. Table 3.7 (Maidment, 1993) lists representative values of both permeability and hydraulic conductivity for various sediment or rock types.

Table 3.7 Representative Values of Hydraulic Conductivity and Permeability

Sediment or Rock Type	Hydraulic Conductivity m/day	Permeability m^2
Clays	10E–7 – 10E–3	10E–19 – 10E–15
Silts	10E–4 – 10E+0	10E–16 – 10E–12
Fine to Coarse Sands	10E–2 – 10E+3	10E–14 – 10E–9
Gravels	10E+2 – 10E+5	10E–10 – 10E–7
Shale (Matrix)	10E–8 – 10E–4	10E–20 – 10E–16
Shale (Fractured & Weathered)	10E–4 – 10E+0	10E–16 – 10E–12
Sandstones (Well Cemented)	10E–5 – 10E+2	10E–17 – 10E–14
Sandstones (Friable)	10E–3 – 10E–0	10E–15 – 10E–12
Salt	10E–10 – 10E–8	10E–22 – 10E–20
Anhydrite	10E–7 – 10E–6	10E–19 – 10E–18
Unfractured Igneous & Metamorphic Rocks	10E–9 – 10E–5	10E–21 – 10E–17
Fractured Igneous & Metamorphic Rocks	10E–5 – 10E–1	10E–17 – 10E–13

From Maidment, D., *Handbook of Hydrology*, McGraw-Hill, New York, 1993. With permission.

3.4.3.6 Hydraulic Conductivity

Hydraulic conductivity is a measure of the capacity of a soil medium to transmit flow of a specific fluid. Unlike permeability, the hydraulic conductivity of a soil type is a function of both the soil medium and the particular fluid. In remediation engineering, the hydraulic conductivity of water is the only value of concern. The term coefficient of permeability is sometimes used in place of hydraulic conductivity, mostly by geotechnical engineers. Hydraulic conductivity is dependent on several factors including pore size and grain size distributions, porosity, and the physical properties of the soil.

Hydraulic conductivity is defined by the formula

$$K = k\mu/\rho g$$

where k = permeability
K = hydraulic conductivity
ρ = density of the fluid
g = acceleration of gravity
m = dynamic viscosity of the fluid

Hydraulic conductivity has the same dimensions as velocity [L/T]. It is usually expressed in the following units: cm/s, m/day, and gpd/ft^2. Table 3.7 (Maidment, 1993) and Figure 3.6 (Heath, 1982) depict some typical values of hydraulic conductivity for different soil materials.

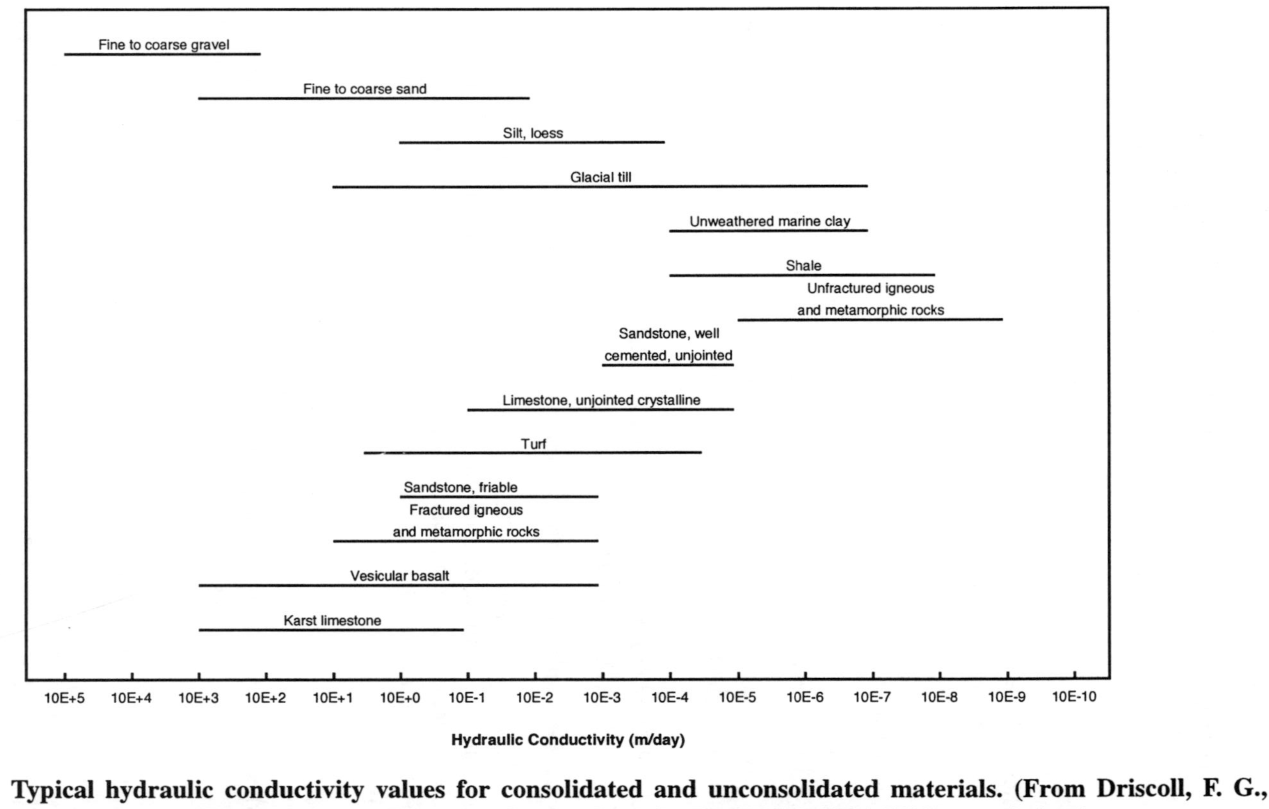

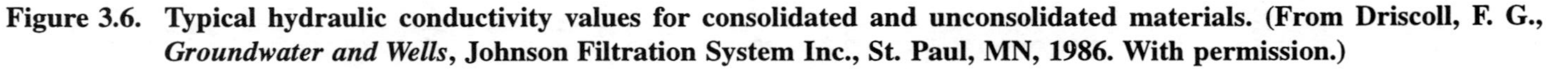

Figure 3.6. Typical hydraulic conductivity values for consolidated and unconsolidated materials. (From Driscoll, F. G., *Groundwater and Wells*, Johnson Filtration System Inc., St. Paul, MN, 1986. With permission.)

3.4.3.7 Transmissivity

The capacity of an aquifer to transmit water is referred to as its transmissivity. Transmissivity of an aquifer is the hydraulic conductivity of the aquifer multiplied by its saturated thickness. It is represented by the formula

$$T = Kb$$

where T = transmissivity
K = hydraulic conductivity
b = aquifer thickness

Transmissivity has the dimensions $[L^2/T]$.

3.4.3.8 Homogeneity and Heterogeneity

An aquifer is said to be homogeneous if its hydraulic properties are the same throughout; if it varies, it is described as heterogeneous.

3.5 FUNDAMENTAL PRINCIPLES OF GROUNDWATER FLOW

3.5.1 Groundwater Flow Velocity—Darcy's Law

Darcy's Law assumes that groundwater flow is laminar. It means that the water will follow distinct flow lines. Since most groundwater flow in porous media is laminar, Darcy's Law is valid for flow of liquid through porous media. The Darcy's Law for groundwater flow is expressed in the equation

$$v = Ki$$

where v = specific discharge or discharge velocity, which is the quantity of water flowing in unit time through a unit cross-sectional area of the medium (flowrate per unit area)
K = hydraulic conductivity
i = hydraulic gradient of the aquifer

Specific discharge has the dimensions [L/T]; however, it is not a measure of the velocity of the groundwater. It is based on the gross cross-sectional area of the soil. The actual velocity of the groundwater through the soil pores is given by the formula

$$v_s = v/n$$

where v_s = pore water velocity or seepage velocity
v = specific discharge
n = porosity

The dimensions of the pore water velocity is the same as for velocity [L/T]. Since $0\% \le n \le 100\%$, it follows that the seepage velocity is always greater than the specific discharge.

Example: The difference in water level in two wells 1 mile apart is 10 ft and the porosity of the soil is 30% with a hydraulic conductivity of 500 gpd/ft². What is the specific discharge and seepage velocity?

$$v = Ki$$

$$= 500\ \text{gpd/ft}^2 \times (10\ \text{ft/5280 ft})$$

$$= 0.95\ \text{gpd/ft}^2$$

$$v_s = v/n$$

$$= \left(0.95\ \text{gpd/ft}^2\right)/0.3$$

$$= 3.17\ \text{gpd/ft}^2$$

In groundwater remediation, the amount of water that a specific formation can produce (also known as yield) is an important piece of information for remediation system design and equipment specification. The quantity of flow per unit time is given by the following formula

$$Q = vA$$

where Q = quantity of flow per unit time
v = seepage velocity
A = cross-sectional area which the flow moves

Example: Continuing from the previous example, given a formation that is 2 miles wide with an aquifer thickness of 20 ft, what is the yield of that formation?

$$Q = vA$$

$$= \left(3.17\ \text{gpd/ft}^2\right) \times (2\ \text{mi} \times 5280\ \text{ft/mi} \times 20\ \text{ft})$$

$$= 669,500\ \text{gpd}$$

3.5.2 Flow Between Aquifers

When there is flow across two aquifers through a confining unit, a modified Darcy's Law can be used to determine the quantity of leakage. The modified Darcy's Law is given by the following formula

$$Q = (p/m)A\Delta h$$

where
- Q = quantity of leakage
- p = vertical hydraulic conductivity of the confining unit
- m = confining unit thickness
- A = cross-sectional area
- Δh = difference in head between the two recovery wells

Example: A 10-ft thick silty confining unit with a vertical hydraulic conductivity of 1 gpd/ft^2 is separating two aquifers. The difference in head between two wells that are tapped into the upper and lower aquifer is 20 ft. Assume these conditions exist in an area of 1/2 square mile. What is the quantity of leakage from the shallow aquifer to the deep aquifer?

$$Q = (p/m)A\Delta h$$

$$= \left(1\ \text{gpd/ft}^2 / 10\ \text{ft}\right) \times \left(1/2\ \text{mi} \times 5280\ \text{ft/mi}\right)^2 \times 20\ \text{ft}$$

$$= 13{,}939{,}200\ \text{gpd}$$

3.6 CONTAMINANT FATE AND TRANSPORT

3.6.1 Contaminant Migration

When a contaminant is introduced at the ground surface, it migrates through the unsaturated zone toward the aquifer. Hence, contaminants move through soil, sediment, fractured rock, human-made conduits, or other pathways on their way to the saturated zone. Migration of LNAPL contaminants can be divided into three stages as follows

- Seepage through the vadose zone
- Spreading over the water table
- Stabilization within the capillary zone

When a significant amount of contaminant is released, it generally migrates downward under the influence of gravity until it reaches the capillary fringe just above the water table. When the contaminant reaches the water table, it spreads over the water table and starts to accumulate on top of the water surface. Ideally, the downward movement of the contaminant will have minimal amount of lateral spreading. However, since the subsurface is usually composed of heterogeneous anisotropic materials, lateral spreading of contaminants should be expected.

In addition to the subsurface material, rate of release and volume of release are the other two major factors affecting the degree of lateral spreading of contaminants in the subsurface. A large instantaneous release into the vadose zone will have a higher degree of spreading in comparison to a continuous small release. Under some circumstances, such as fingering (depicted in Figure 3.7), a continuous small release will also have a significant degree of spreading.

3.6.2 Volatilization

After contaminants are released into the subsurface, part of the contaminants will volatilize into the vapor phase and move by dispersion from areas of higher concentration to areas of lower concentration. Eventually, an equilibrium between the liquid phase and vapor phase of the contaminant will be established near the source of contamination. Air movement within the subsurface will disturb the equilibrium and promote volatilization; however, the movement of air in the vadose zone is very slow under natural conditions. In soil remediation, soil vapor extraction is effective in removing contaminants from the subsurface by greatly increasing the air movement in the subsurface.

The fate of volatilized vapors in the vadose zone includes releasing to the atmosphere, adsorption by soil particles, consumption by bioactivities, or redissolving in moisture trapped in soil pores.

3.6.3 Adsorption

Adsorption is a phenomenon where molecules of the contaminants dissolved in water attach themselves to the surface of an individual soil particle. This surface attachment can be physical adsorption, chemical adsorption, or exchange adsorption. Physical adsorption is caused by Van der Waals forces and is the easiest to separate. Chemical adsorption is formed by chemical bonding and requires significant efforts to separate. Exchange adsorption is caused by the electrical attraction between adsorbate and the surface.

Adsorptive reactions of organic contaminants moving through soil are typically of the chemical type. A concentration equilibrium will eventually be reached between the contaminant molecule dissolved in the liquid phase and the contaminant molecule that is attached to the soil particles. When concentration conditions change, the soil may adsorb more organic molecules or release them. Since adsorption is a surface phenomenon, its activity is a direct function of the surface area of the solid and the electrical forces on that surface.

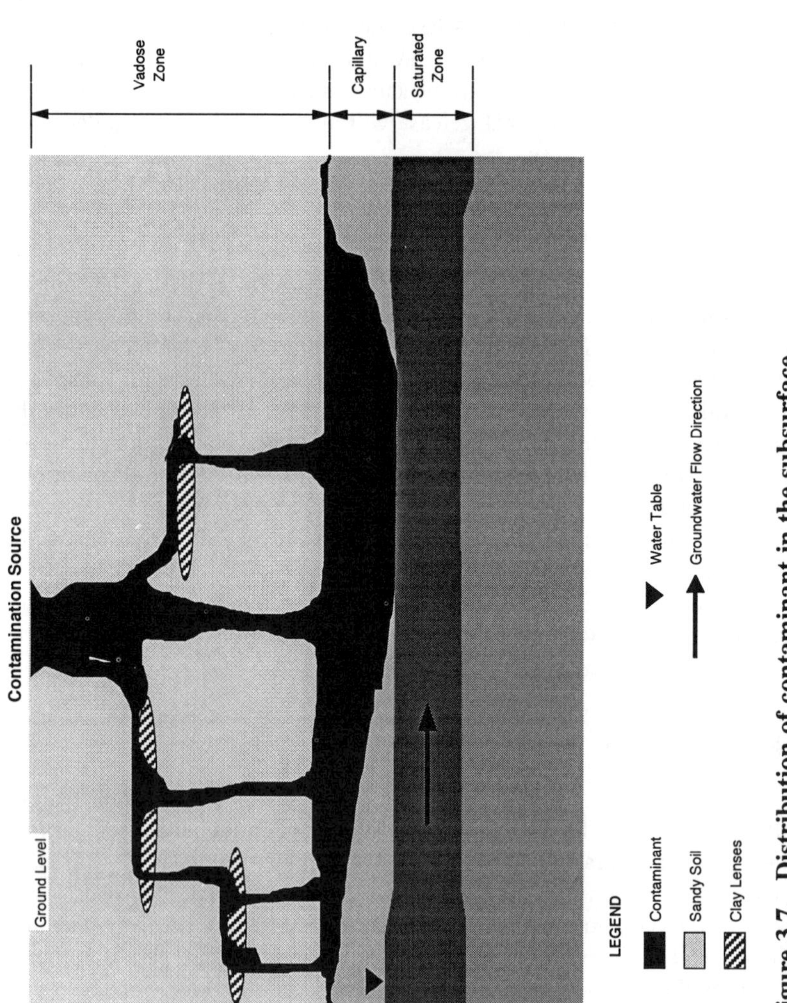

Figure 3.7. Distribution of contaminant in the subsurface.

3.6.4 Contaminant Pathways Through the Vadose Zone

Releases from buried tanks, pipelines, building basements, or any subsurface source can promote the migration of contaminants, since these locations are below the surface where direct observation is not possible, and typically, the releases are not realized for a long period of time. These releases place material directly in the subsurface vadose zone and will eventually migrate both vertically and laterally down to the groundwater. If the groundwater is shallow, the transport time will be decreased significantly.

Contaminant movement through the vadose zone will tend to follow fluid migration. If cracks or fractures exist, the contaminant may flow directly into the crack and move downward. Movement through the subsurface may be enhanced by biological structures and voids in soil or sediment. Although these voids may only be tenths of an inch in diameter, they can transmit fluid in both horizontal and vertical directions.

3.6.5 Contaminant Movement Within the Aquifer

Contaminant movement within the aquifer depends mainly on the properties of the contaminant, specifically the density of the contaminant. Three general migration pathways commonly investigated are floaters, mixers, and sinkers. These pathways are depicted in Figure 3.8. Floaters are the immiscible contaminants that float on top of the water table because they are less dense than water, mixers are the contaminants with uniform dissolution and movement in the aquifer, and sinkers are contaminants that move vertically due to higher density makeup.

These general conceptual pathways are grossly simplified depictions of subsurface contamination movement scenarios; however, as with all real world situations, the subsurface reality is much more complex. Besides density, other parameters such as chemical composition, molecular weight, solubility, viscosity, groundwater velocity, and aquifer geology will also influence contaminant movement.

Dissolved contaminants are transported by both advective and diffusive mechanisms. Advection is the principal mode of transport and results from the hydraulic gradients present in the saturated zone. In the absence of significant advective movements (due to hydraulic gradients that are very small), transport of dissolved contaminants still occurs by diffusion if concentration gradients are present.

3.6.6 Effects of Aquifer Stratigraphy on Contaminant Movement

The presence of sand and clay will profoundly affect the movement of the contaminant in the aquifer. For example, the presence of thin clay lenses

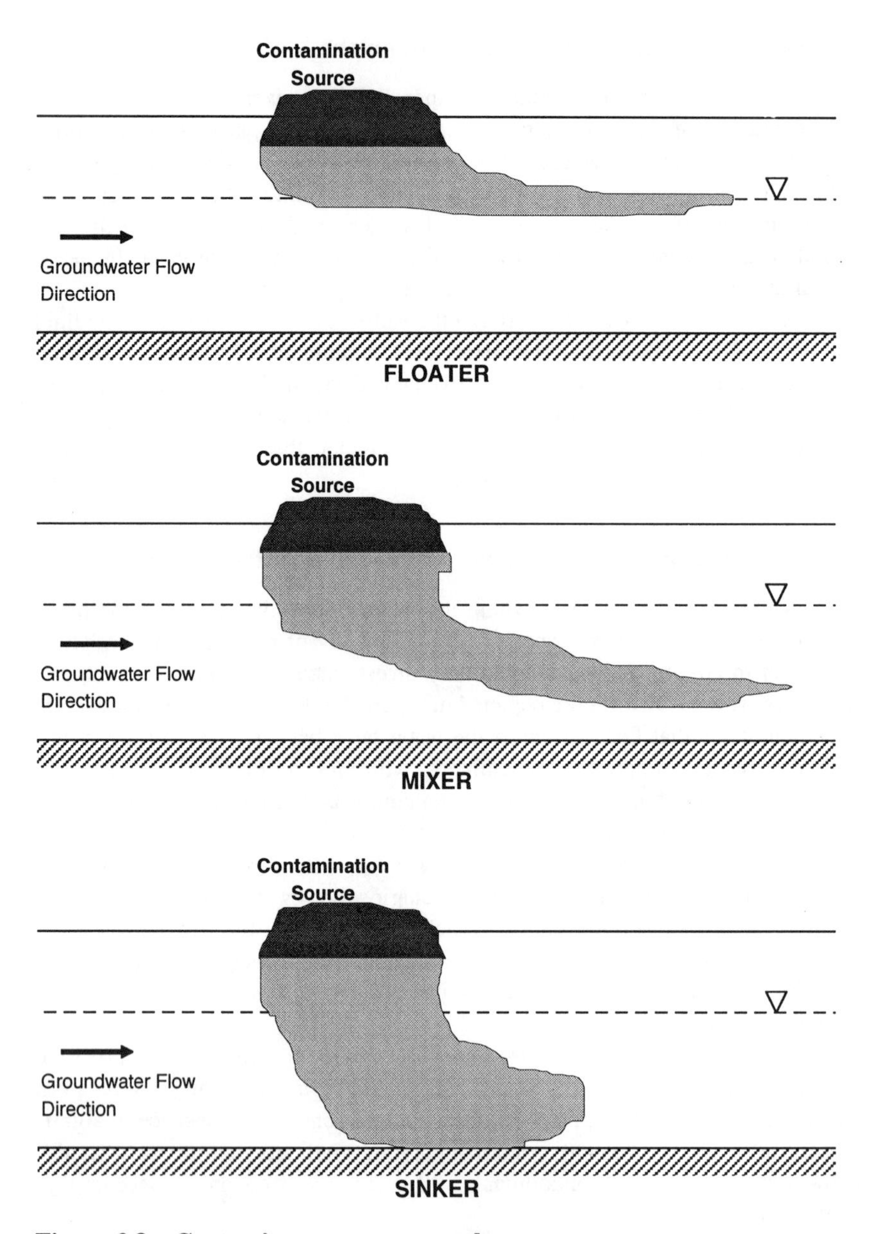

Figure 3.8. Contaminant transport pathways.

in the aquifer may split, retard, or deflect flow, causing the plume to spread horizontally and vertically. Similarly, if the aquifer is filled with fine-grained materials, the plume may move slowly at lower flow rates than calculations may imply. On the other hand, the presence of sand lenses in the formation will allow for more rapid contaminant migration.

3.6.7 Contaminant Movement Between Aquifers Through Aquitards

Aquitards that separate individual aquifers have very low hydraulic conductivities and are impermeable. In reality, aquitards transmit fluid at a very slow rate. When contaminants enter the subsurface system, they may move as a natural chemical system would. However, movement can be enhanced since the materials can be introduced at any depth and in any geological unit.

The most common case is through a well, typically an irrigation well, that taps several aquifers for maximum yield, connecting several aquifers that enable groundwater movement between aquifers. When several aquifers are interconnected via a pumping well, the pumping action that draws the contaminant toward the well will eventually introduce the contaminant to the noncontaminated aquifer.

4 DESIGN APPROACH

4.1 INTRODUCTION

It would not be too far from the truth to say that the average remediation design engineer has heard the phrase "design of remediation systems is not exactly rocket science" many times over. While probably not intended as a downplaying of the role of engineers in the design of treatment systems, it does reflect the attitude and belief of some in the industry that remediation design is, or at least should be, a fairly simple and straight-forward process. After all, it has at times been labeled "catalog engineering" and many installations, especially those at small corner service station-type sites are supposedly "boiler plate-type" designs. Additionally, it is pointed out that the concepts behind widely used remedial techniques have been well-established in various disciplines of engineering. For example, soil venting and air stripping mechanisms are well-documented in chemical engineering. Bioremediation borrows extensively from the knowledge and technology advanced in sanitary or wastewater engineering. The aboveground treatment portion of pump and treat technology comes almost exclusively from the discipline of water supply engineering.

However, while this is all true, clear evidence has been presented that "borrowed technology and off-the-shelf equipment" does not necessarily equate to a successful system. On the contrary, many remedial design engineers have had frustrating experiences with above-ground treatment units completely plugged by precipitates and bacteria, dual-pump systems that do not remove product, frozen air stripping towers with "icebergs" in them and spewing out water vapor, and thermal/catalytic oxidizers with unacceptably high downtimes.

The key to designing successful remediation systems is twofold: careful selection and application of the appropriate technology (based on understanding of site-specific conditions), and designing with the operation of the system in mind (taking into consideration construction aspects and operations and maintenance issues). In order to accomplish this, the engineer should approach the design with the following considerations:

- thorough knowledge and understanding of the site: contaminant type, geology, hydrogeology, site restrictions, etc.
- an accurately surveyed site plan
- design concept
- likely operations and maintenance problems that will be encountered
- foreseeable public relations problems: noise, odor, etc.
- regulatory concerns
- health and safety issues
- effectiveness of system in terms of attainable cleanup, cost, and time
- preparation of site-specific plans and specifications
- construction involvement
- preparation of accurate as-built plans
- preparation of monitoring/sampling plans and operations and maintenance manual

These considerations can be incorporated into a systematic design process, outlined in the next section.

4.2 TYPICAL DESIGN PROCESS

When a contaminated site is discovered, a detailed site assessment investigation will be initiated. The purpose of the study is to determine if groundwater is impacted and to delineate the vertical and lateral boundaries of the contamination. A report detailing the findings and conclusions of the investigation will be generated and is usually referred to as a Detailed Hydrogeological Assessment Report or a Problem Assessment Report (PAR). While the team geologist or hydrogeologist will be heading up the investigation, the design engineer should be involved somewhat. On-site drilling usually involves boreholes (temporary) and monitoring wells (permanent). Often, some of the monitoring wells will later be converted to remediation wells in the design stage. As the assessment proceeds, an engineer's input will certainly prove useful. The PAR should be as extensive and conclusive as possible because the ultimate design will be based on it.

Based on the assessment's findings and regulatory requirements, the engineer will evaluate several design alternatives to select the most appropriate technology. This should be a carefully thought-out process. A Remedial Action Plan (RAP) report or Corrective Action Plan report will then be generated. This report examines the site conditions detailed in the PAR and evaluates various remedial alternatives. Following the RAP, a pilot test is usually performed. Sometimes the RAP is performed after the pilot test is completed. Based on the results of the pilot test, the final remediation design will be performed. Sometimes a final remediation plan (FRP) will then be prepared.

The following summarizes a typical chronology of activities leading to final design of a remedial system.

1. Detailed Hydrogeologic Site Assessment or Problem Assessment Report—delineating horizontal and vertical extent of contamination; determining site geologic and hydrogeologic conditions.
2. Cleanup Goals/Regulatory Requirements—liaison with regulatory agencies.
3. Remedial Action Plan or Corrective Action Plan—evaluates treatment options and selects optimum remedial plan.
4. Feasibility Study or Pilot Test—conduct pilot test to establish feasibility and design parameters.
5. System Design—engineering design of remedial system.

As the purpose of this book is to serve as a guide for the design of the remediation system itself, the procedures for performing the pilot tests and system designs will be addressed. Generally, after the pilot tests have been concluded, the system design itself typically involves the following four stages:

(1) System Planning
  Define basis of design
  - Summarize previous findings and conclusions
  - Establish design objectives
  - Identify design parameters

  Perform preliminary design
  - Site information gathering
  - Site survey
  - Preliminary equipment list
  - Sketch of system layout
  - Estimate cost of system

(2) System Design
  Conceptual Design
  - Review preliminary design
  - Finalize design concept
  - Prepare appropriate process instrumentation and control diagrams

  Final Design
  - Perform detailed design evaluations as needed
  - Prepare detailed plans and specifications to describe and locate proposed remedial system in the contract documents

(3) System Construction
  Bidding Process
  - Receive and evaluate bids and select bidder

  Construction Management
  - Provide resident inspection
  - Provide construction administration services

(4) System Operation
  - Prepare operation and maintenance manual
  - System start-up

The first stage of the design process is planning. The importance of this cannot be stressed enough. Good system planning provides the foundation for the remainder of the design process. System planning includes preparation of a synopsis of all previous findings and conclusions, and performing a preliminary design. There are generally three steps involved in the preparation of a synopsis: defining current conditions and problems, establishing design objectives, and identifying design parameters.

In performing the first step, i.e., defining site conditions and problems, particular attention should be paid to the following subjects:

- Site geological characteristics: type of soils, cross-sectional profiles of subsurface
- Site hydrogeological conditions: water table depth, seasonal fluctuations of water table, subsurface transport mechanism
- Contaminant information: types of contaminants, concentration of contaminants, locations of contaminants, vertical and horizontal delineation of contaminants, source(s) of contaminants
- Site conditions: type of location, neighboring properties
- Systems or system components installed to-date: types of wells, well logs, remediation systems
- Feasibility studies or pilot tests: types of test performed, test methods, results of tests, conclusions of tests, adequacy of tests

This process involves a methodological review and analysis of relevant existing reports, including detailed hydrogeological reports or problem assessment reports, remedial action plans or corrective action plans, pilot test results, and the latest quarterly monitoring reports.

The second step is to define the goals for managing the site. Is containment of the plume (i.e., preventing the plume from spreading) or cleanup (i.e., meeting specific regulatory levels for allowable contaminant concentrations in the subsurface) the objective of the remedial action? Although these issues would probably have been discussed in the RAP, additional information from feasibility tests would certainly throw more light onto the subject. Based on the pilot test results, answers to the following questions will be more obvious.

- Is vadose zone remediation feasible?
- Is groundwater contamination cleanup feasible?
- What are the required cleanup goals?
- Which remedial option(s) should be used?
- What is the estimated time-frame for cleanup?

It is important to clearly state the objectives of the remedial system with regards to the restoration of the site. During operations, the performance and the efficiency of the system will be closely scrutinized. Thus, it is good practice to define the objectives and goals of the treatment system in the conceptual design report.

The third step is to establish design criteria that would form the basis for the design process. These parameters and assumptions will be used in all design calculations and treatment equipment selection.

The preliminary design stage typically involves site data and information gathering and formulating a remedial system outline. Major components of the overall system are selected at this stage.

The actual design can be arbitrarily divided into the conceptual design and the final detailed design. The conceptual design generally involves a review of the preliminary design (with client and/or regulators), finalizing the design concept, and preparing the piping and instrumentation diagram. In the final detailed design process, engineering calculations are performed and engineering drawings and technical specifications are prepared. For a typical remediation system design, engineering calculations generally involve head loss calculations, pump/blower size determinations, life-cycle analyses, treatment equipment sizing, etc.

4.3 RISK-BASED CORRECTIVE ACTION

Traditionally, the ultimate goal of a groundwater remediation system is to clean up contaminated groundwater to background levels or to a set of very low criteria set by the regulatory agencies, such as maximum contaminant levels (MCL). These were levels that were established in the interest of protecting human health and the environment, and have consequently resulted in profuse financial expenditures across the country. In many cases, expenses for corrective action could be reduced substantially without jeopardizing the environment. As a result, these cleanup levels are being questioned and reevaluated to determine "how clean is clean."

A strategy called Risk-Based Corrective Action (RBCA, pronounced "Rebecca") for petroleum release sites was introduced in response to needs expressed by both industry and regulatory agencies. Since then, the American Society of Testing Materials (ASTM) has developed a standardized approach to risk-based corrective action for petroleum release sites (ASTM Method E1793-95). Even though ASTM was credited for developing the standard method, the process of developing RBCA involved many parties, including the petroleum industry, government agencies, and insurance, banking, and environmental consulting firms. While the RBCA process is not restricted to certain contaminants, the ASTM guide emphasizes the application of RBCA to petroleum fuel releases.

The two main goals of RBCA are to achieve practical and cost-effective cleanup of contaminants and to protect human health and the environment. The ASTM guide is a general framework for conducting RBCA. It was written in a way that state and local programs can customize the framework to be consistent with legislation, policy, and the needs of each individual agency.

Specifically, the agencies involved will need to develop a classification system and a Tier 1 Look-Up Table. The use of a RBCA framework has been proven to help improve project management for both the regulated community and the regulator.

RBCA is based on a multitiered approach for achieving closure or no further action status and integrates U.S. Environmental Protection Agency risk and exposure assessment guidelines. The typical flow in a RBCA process is from site classification (based on the urgency of initial responses) to implementation of appropriate initial response actions to the development of target correction action levels to developing a corrective action plan. It is an approach in which risk and exposure assessment practices are integrated into traditional components of the corrective action process to ensure that appropriate and cost-effective remedies are selected and that limited resources are properly allocated (see Figure 4.1).

The RBCA process consists of 3 tiers (see Figures 4.2, 4.3, and 4.4). Tier 1 is designed as a generic screening level that helps focus site assessment and additional risk evaluation, if necessary. Tier 1 process is considered a conservative and efficient method. Under Tier 1, a site is classified based on information collected from visual inspection, historical records, and minimal site assessment data. Typically, a Tier 1 site assessment will involve the identification of contaminant sources, obvious environmental impacts, the presence of potentially impacted human and environmental resources (for example, residents, surface and underground water bodies, etc.), and potential significant transport pathways (for example, surface water flow, groundwater flow, atmospheric dispersion, etc.).

In addition, part of the Tier 1 protocol includes a "Look-Up Table" for screening concentrations to determine if site conditions satisfy a quick regulatory closure or require additional site-specific evaluation of corrective action goals. It is advised that before using this table, the appropriate regulatory agencies be contacted to ensure that the appropriate values are used in the risk calculation. This is due to the fact that these values may change as new methodologies and parameters are developed.

Tier 2 and Tier 3 processes are based on site-specific information and require additional resources. Obviously, the cost of regulating, managing the remediation site, and collecting site-specific information will increase from Tier 1 to Tier 2 and from Tier 2 to Tier 3. In most cases, the decision to move to a higher tier is based on answers to the following questions:

- Do the assumptions used in the lower tier appropriately or reasonably reflect the conditions of the site?
- Are the costs for the targets established from a higher tier analysis lower than those established in a lower tier analysis?
- Is it more costly to achieve the lower tier's target or to conduct additional analyses for the higher tier?

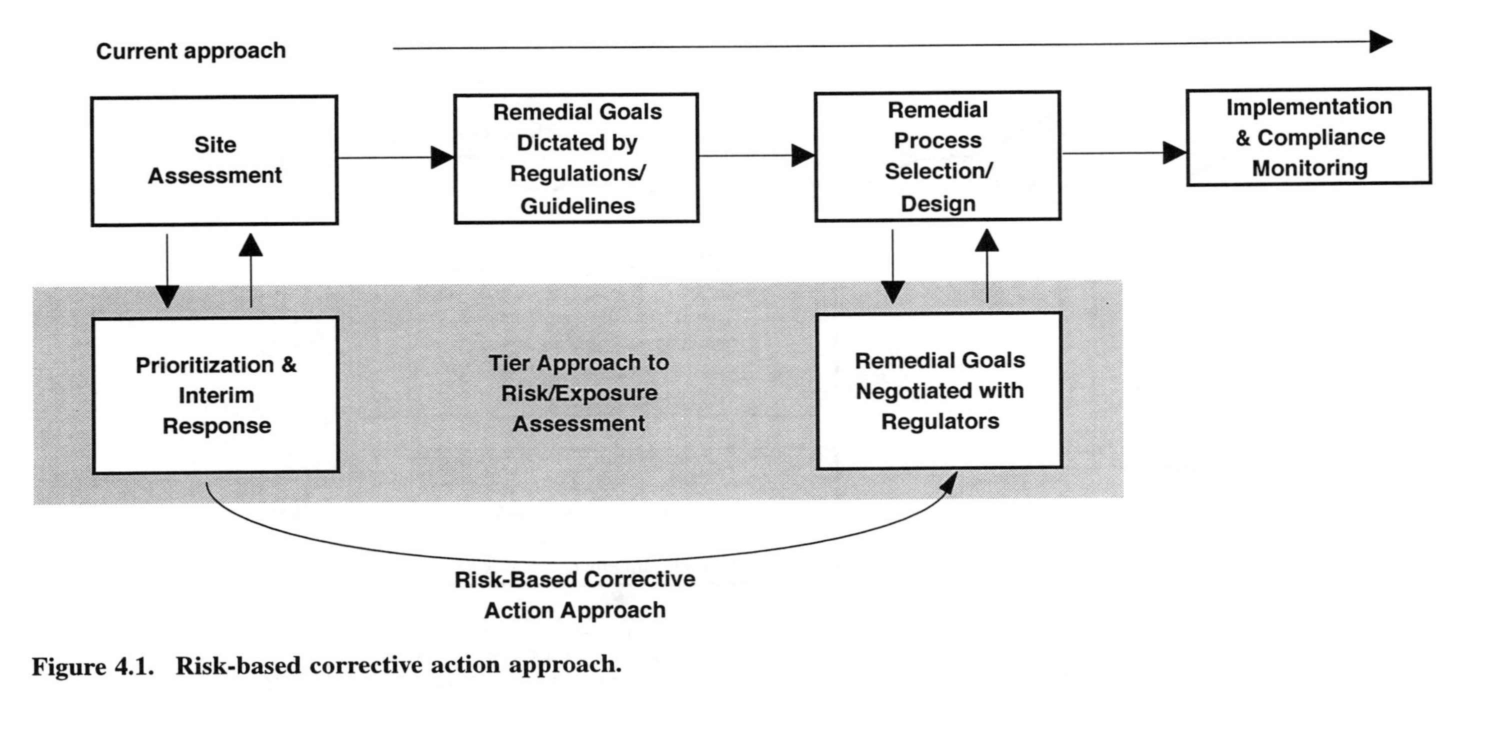

Figure 4.1. Risk-based corrective action approach.

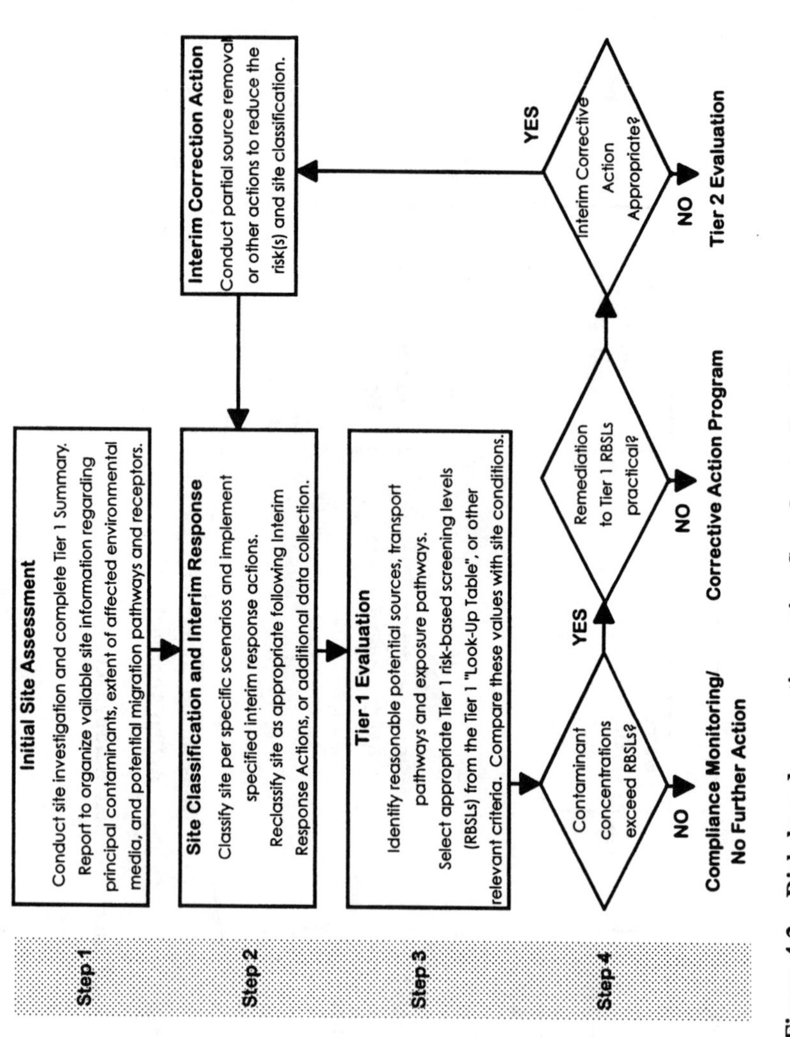

Figure 4.2. Risk-based corrective action flowchart—Part I.

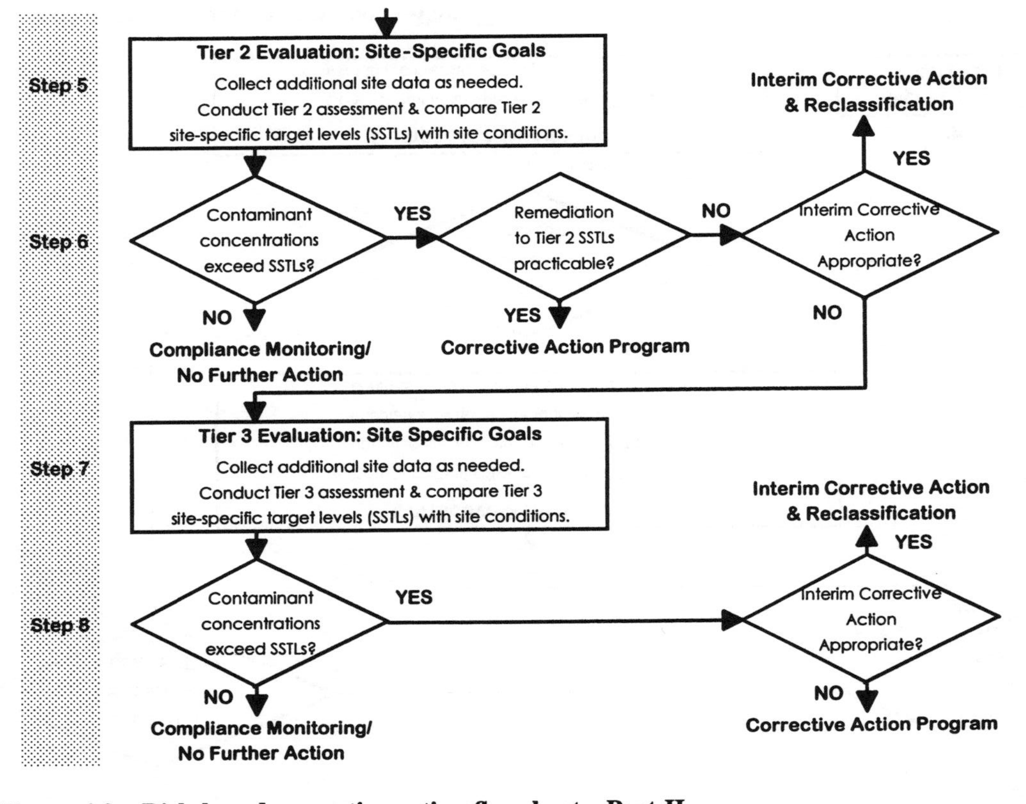

Figure 4.3. Risk-based corrective action flowchart—Part II.

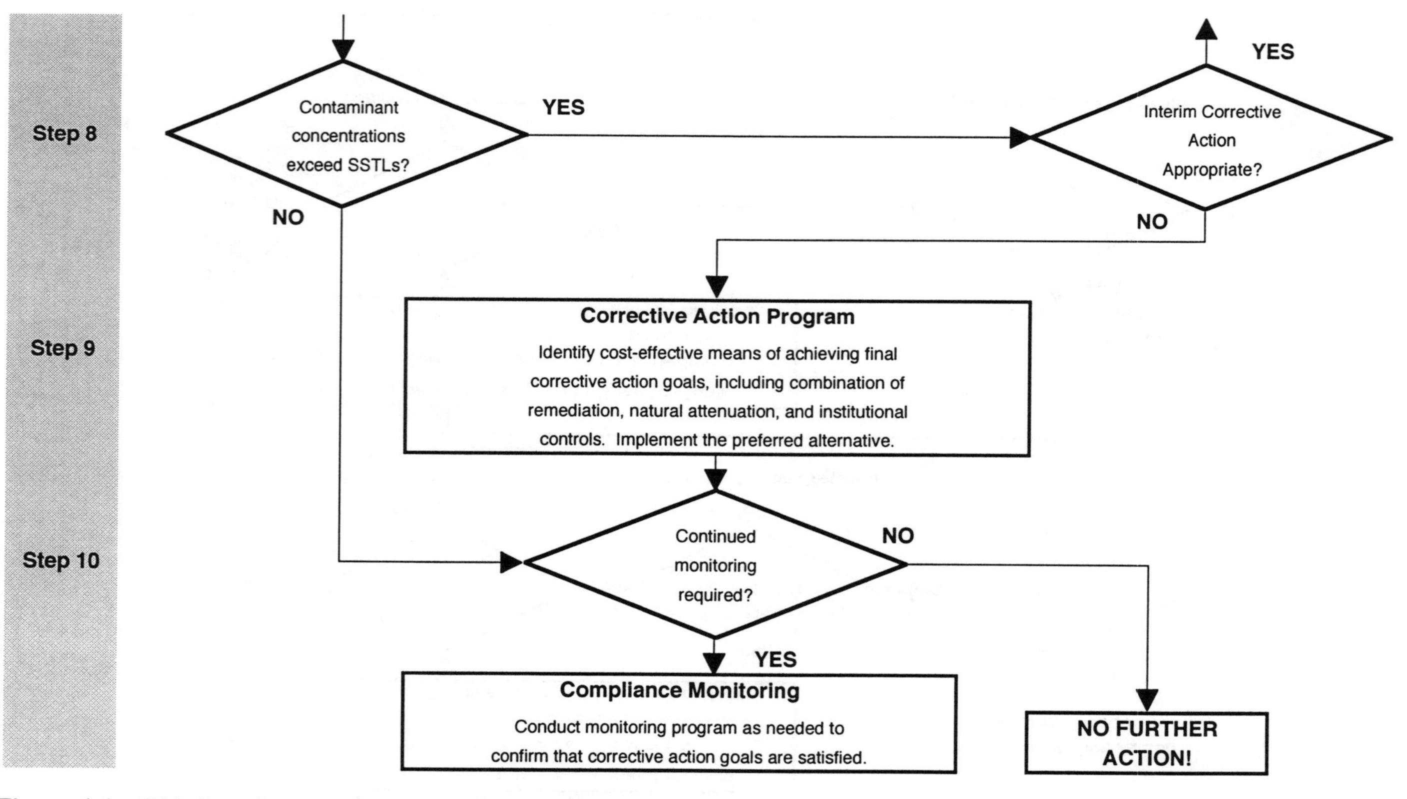

Figure 4.4. Risk-based corrective action flowchart—Part III.

The difference between Tier 1 and higher tiers is that in a higher tier the user(s) can develop more cost-effective corrective action plans by determining site-specific target levels (SSTL) based on site-specific and realistic information, as opposed to relying on the conservative assumptions presented in the Tier 1 process.

While both Tier 2 and Tier 3 involve developing SSTL, the main distinction between Tiers 2 and 3 is that Tier 2 analyses tend to be consistent with the level of site characterization data most often available, and Tier 3 often involves a much more significant increase in site-specific data requirements and the use of complex numerical models and probabilistic analyses such as the Monte Carlo Simulation.

Software tools currently available in the market that can be used as part of the RBCA process include API DSS and CALTOX. The level of model sophistication should be commensurate with the appropriate tier. In some states, site classification will determine if direct government agency oversight is needed and the corrective action may be self-directed or conducted under the oversight of "licensed site professionals."

5 DESIGN OF SOIL VAPOR EXTRACTION SYSTEMS

5.1 INTRODUCTION

Soil vapor extraction (SVE), also known as vacuum extraction, enhanced volatilization, in situ volatilization, soil venting, and in situ aeration, is a relatively low-cost and effective soil remediation technology for soils contaminated with volatile organic compounds (VOC) and petroleum hydrocarbons. A typical SVE system includes vapor extraction wells, blower, condensate separator (also known as knockout tank), control valves, pressure gauges, and flow meters. If vapor abatement is required, vapor treatment methods such as catalytic oxidation, thermal destruction, or carbon adsorption of extracted vapor can be employed. A simplified SVE system with vapor treatment is depicted in Figure 5.1.

The application of SVE technology in soil remediation has increased significantly over the years due in part to several reasons, including:

1. SVE can treat large areas of soil contamination with minimum site disturbance.
2. The costs of installing SVE systems are relatively low compared to many other soil remediation options, such as excavation and incineration.
3. SVE effectively reduces the concentration of VOC residue in the vadose zone and, to some extent, free product on the water table, which in turn reduces further contamination of the soil and groundwater due to vapor and free product migration.
4. SVE systems are relatively easy to construct. The components of an SVE system are usually readily available off-the-shelf from equipment vendors.
5. The design and operation of SVE systems, especially for typical corner gas station sites, has become almost a "standard" procedure.

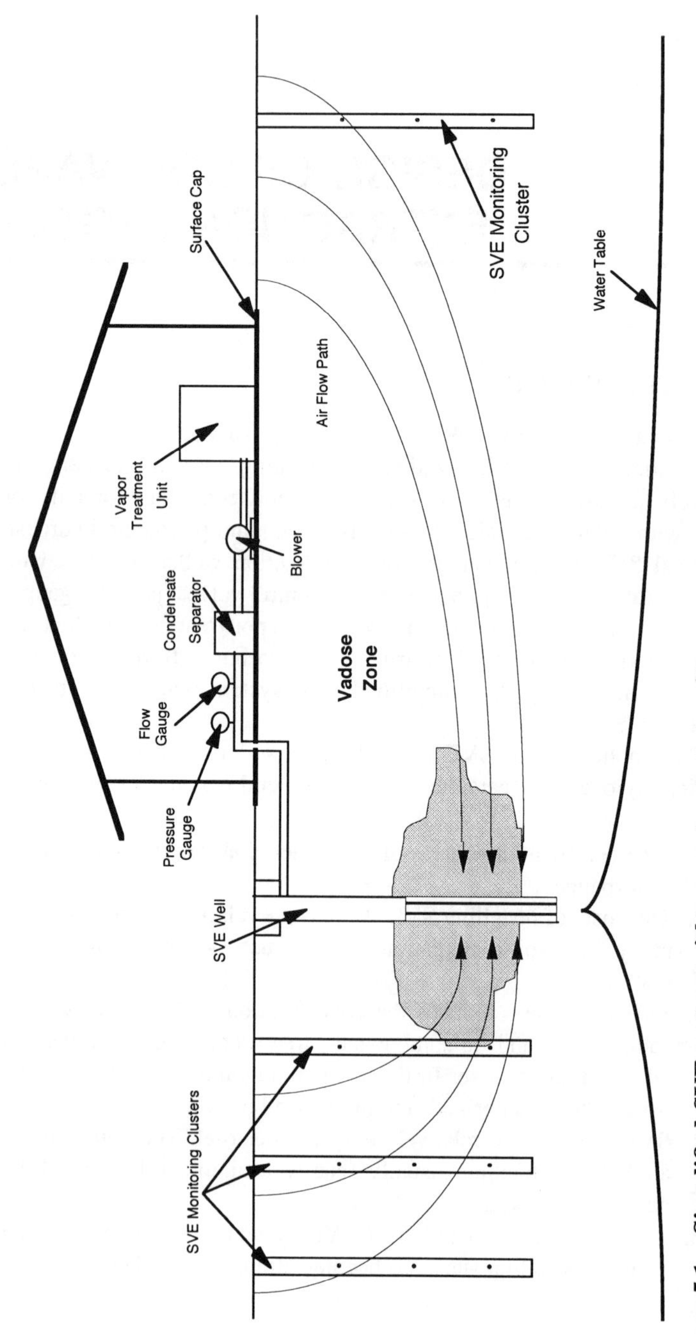

Figure 5.1. Simplified SVE system with vapor treatment unit.

Almost all of the SVE systems in operation throughout the country were installed based on one or more of the aforementioned reasons. Because SVE systems have become almost a standard solution for gasoline spill sites, "cut-and-paste" designs, as a result, are not uncommon. The problem with over-simplifying engineering designs is that it sometimes overlooks critical aspects of the design. The design, operation, and monitoring of SVE systems can be quite complex because of variations in site geology, specific site conditions (presence of building, underground utilities, etc.), type of contaminants, contaminant characteristics, extent of contamination, groundwater elevations, and, more critically, seasonal fluctuations of the water table and regulatory cleanup criteria.

Certainly, there are sites and occasions when conditions are such that minimal efforts should be expended in designing a simple SVE system. Likewise, conducting pilot tests may not even be justified at some of these sites. However, the broad-based application of such "oversimplification" is ill-advised. Extraction well screen intervals and depths are critical for the successful implementation of SVE and should be well thought out. Closing of well screens due to upwelling effects and seasonal fluctuations of the water table are serious considerations, especially if the vertical extent of contamination reaches close to the water table. It is the intent of this chapter to assist the design engineer in evaluating the factors and conditions that will affect the design of an SVE system.

5.2 SOIL VAPOR EXTRACTION CONCEPT

The SVE system operates on the principle that organic hydrocarbons contained in petroleum products can volatize at temperatures typically found in the soil. Volatilization is the transfer of substance from a solid or liquid phase to a vapor phase. Since volatilization forms the basis of SVE, the concept for this technology relies heavily on Henry's law.

In Chapter 3 it was shown that petroleum products in the vadose zone exist in the vapor phase, dissolved phase (in pore water), liquid phase or non-aqueous phase liquids, and adsorbed phase. The SVE system is then based on two mechanisms: introducing fresh air (flux) into the vadose zone to volatize dissolved phase, liquid phase, and adsorbed phase hydrocarbons to the vapor phase (mass transfer), and moving the vapor along with the flux to the extraction well (removal). Thus, the success of an SVE system basically depends on the volatility of the contaminant and the air permeability of the soil.

Many different factors affect the success of an SVE system. In general, these factors can be divided into three categories: characteristics of the contaminant, properties of the soil, and site conditions. The design engineer obviously has little or no control over the first two groups, and somewhat limited control over the third category. A fourth category, the design of the

SVE system itself, presents the engineer with an opportunity to optimize the system for both cost-effectiveness and successful cleanup. The following sections will provide brief descriptions on the various factors that influence SVE systems.

5.2.1 Contaminant Characteristics

Mass transfer is a function of contaminant characteristics because they dictate the ease of volatilization. Volatilization in the vadose zone is the transfer of a compound from a liquid phase to the vapor phase. The volatility of the compound is probably the single most important criteria affecting the applicability of SVE. Volatilization is governed by properties of the compound and the soil. Properties governing volatilization include vapor pressure, water solubility, Henry's law constant, melting point and boiling point, organic carbon partition coefficient, and contaminant composition.

5.2.1.1 Vapor Pressure

Vapor pressure is a measure of the tendency of a substance to pass from a liquid to a vapor state. It is the pressure of the gas in equilibrium with the liquid at a given temperature. The greater the vapor pressure, the more volatile the substance. Temperature affects the vapor pressure of a compound, with the vapor pressure increasing dramatically as the temperature is increased. As such, vapor pressure is always reported with the temperature at which the pressure was measured. The vapor pressure of a contaminant can be estimated from the following relationship:

$$P_v = CH_L \tag{5.1}$$

where P_v = vapor pressure of the contaminant (atm)
C = concentration of the contaminant in water (mole/m^3)
H_L = Henry's law constant (atm-m^3/mole)

5.2.1.2 Water Solubility

Water solubility is an important property of organic substances. For a gas, this must be measured at a given vapor pressure. For a liquid, it is a function of the temperature of the water and the nature of the substance. Solubilities of organic materials can range from completely miscible with water to nearly insoluble. More soluble materials have a greater potential mobility in the environment. Laboratory solubility is measured using distilled water, which may not have the same effect as groundwater. Tables 2.1 and 2.2 list the solubility data for some commonly encountered compounds.

5.2.1.3 *Henry's Law Constant*

Henry's law states that there is a linear relationship between the partial pressure of a gas above a liquid and the mole fraction of the gas dissolved in the liquid. It is given as:

$$H_L = P_x / C_x \tag{5.2}$$

where H_L = Henry's law constant (atm-m³/mole)
P_x = partial pressure of gas at a given temperature (atm)
C_x = equilibrium concentration of the gas in solution (mole/m³)

Henry's law is valid if the gas is soluble, the gas phase is reasonably ideal, the gas will not react with the solute, and the solution is very dilute. Thus, it is very useful when applied to volatile organic compounds that are dissolved in water, such as benzene, toluene, carbon tetrachloride, trichloroethylene, etc.

Compounds with a high Henry's law constant will tend to have a greater concentration in air when an air/water system is in equilibrium. These substances undergo a phase change from the dissolved state to vapor easily, and hence are easily removed from the dissolved phase in the soil with the SVE system. Compounds with a low Henry's law constant are more hydrophilic and are more difficult to strip. Tables 2.1 and 2.2 provides the Henry's law constants for some commonly encountered materials.

5.2.1.4 *Melting Point and Boiling Point*

Melting point and boiling point can be used to evaluate whether a particular compound will be gas, liquid, or solid at a certain temperature. If the specified temperature is below the melting point, the compound will be a solid. If the temperature falls between the melting point and the boiling point, the compound will be a liquid, and if the temperature is above the boiling point, the compound will be a gas. Boiling points are usually given at a pressure of 1 atm (760 mmHg). At a compound's boiling point, its vapor pressure equals the vapor pressure of the atmosphere. Generally, boiling point increases with increased molecular weight.

5.2.1.5 *Organic Carbon Partition Coefficient*

Organic carbon partition coefficient is the ratio of the amount of contaminant adsorbed per unit weight of organic carbon in the soil to the concentration of the chemical in solution at equilibrium. It is expressed as K_{oc}:

$$K_{oc} = \frac{[\text{mass of sorbed compound/mass of organic carbon}]}{[\text{mass of compound in solution/volume of solution}]} \tag{5.3}$$

Table 5.1 Regression Equations for the Estimation of K_{oc}

Equation	Chemical Classes Represented
$\log K_{oc} = -0.55 \log S + 3.64$ (S in mg/L)	Wide variety, mostly pesticides
$\log K_{oc} = -0.54 \log S + 0.44$ (S in mole fraction)	Mostly aromatic or polynuclear aromatics
$\log K_{oc} = -0.557 \log S + 4.277$ (S in micromole/L)	Chlorinated hydrocarbons
$\log K_{oc} = 0.544 \log K_{ow} + 1.377$	Wide variety, mostly pesticides
$\log K_{oc} = 0.937 \log K_{ow} - 0.006$	Mostly aromatic or polynuclear aromatics
$\log K_{oc} = 1.00 \log K_{ow} - 0.21$	Mostly aromatic or polynuclear aromatics
$\log K_{oc} = 1.029 \log K_{ow} - 0.18$	Variety of insecticides, herbicides, and fungicides

Sorption of contaminant liquids to soil particles and organic matter is a very important factor. It is a phase distribution process in which dissolved organic compounds are transferred from the aqueous phase to the solid phase. Sorption is a major mechanism affecting the mobility of dissolved organic compounds. The sorption of dissolved organic compounds is controlled primarily by the organic matter components of soils. The contributions of inorganic components, such as clay minerals, are relatively small.

This relationship is valid even if the organic carbon content of the soil is very low. The organic carbon partition coefficient is related to the distribution or sorption coefficient, K_d, as follow:

$$K_{oc} = K_d / f_{oc} \tag{5.4}$$

where f_{oc} is the fraction of organic carbon associated with the soil.

Many correlations have been developed to accurately predict the K_{oc} for a particular chemical compound. Typically, these correlations will involve the aqueous solubility (S) or octanol–water partition coefficient (K_{ow}) of the compound. With knowledge of S and K_{ow}, which are available for most compounds, the K_{oc} for a particular compound can be estimated. Table 5.1 lists some of the correlations expressed in terms of S, K_{oc}, and K_{ow}, and Table 5.2 lists the octanol–water coefficient for some common hydrocarbon constituents.

5.2.1.6 Contaminant Composition

The most common products released from underground storage tanks are petroleum hydrocarbons such as gasoline and diesel fuel. The compo-

Table 5.2 Octanol–Water Coefficient for Selected Hydrocarbon Constituents

Compounds	K_{ow}
Acrylonitrile	1.38
Alachlor	830.00
Aldicarb	11.02
Atrazine	476.00
Benzene	135.00
Biphenyl	7,540.00
Bromobenzene	900.00
Bromodichloromethane	76.00
Bromoform	200.00
Bromomethane	12.59
Chlorobenzene	692.00
Chloroethane	34.67
Chloromethane	8.00
Dibromochloromethane	123.00
1,2-Dichlorobenzene	2,399.00
1.3-Dichlorobenzene	2,455.00
Dichlorodifluoromethane	144.54
2,4-Dichlorophenol	562.00
2,4-Dinitrophenol	34.00
2,4-Dinitrotoluene	102.00
Ethylbenzene	1,413.00
Malathion	780.00
Methylene Chloride	17.78
Naphthalene	2,040.00
Phenol	1.46
1,1,1-Trichloroethane	148.00
Trichloroethylene	195.00
1,1,2,2-Tetrachloroethane	363.00
Tetrachloroethylene	758.00
Toluene	490.00
Trichlorofluoromethane	338.84

sition of gasoline released into the subsurface will change with time due to the volatilization of the more volatile compounds and the biodegradation of some of the gasoline compounds by the indigenous microbial population present in the subsurface. The application of SVE will remove the more volatile compounds rather quickly and its effectiveness will drop significantly when treatment is down to the removal of the heavier low-volatility compounds.

5.2.2 Soil Properties

Mass transport or extraction is a function of soil properties, because they dictate how easily the flux will move through the vadose zone. Thus, the feasibility and design of SVE systems are essentially based on site-specific soil conditions. Soil properties that are critical to the success of SVE systems include permeability and moisture content.

5.2.2.1 Air Permeability

The air permeability of a soil is simply a measure of the ability of vapors to flow through porous media. It is the most important parameter in the success of SVE technology and is a key parameter in deciding if SVE is a feasible remedial option. In addition, it also plays a significant part in the SVE system design specifically in determining the number of extraction wells required to remediate a site.

The air-filled pore space in the subsurface is the basic determinant of the volume available for vapor transport. The size of the pore space and its interconnections what is conducive toward good air permeability. Presence of water in the subsurface greatly affects the vapor transport mechanism because it takes up pore spaces that are available for vapor transport. In addition to the moisture content in the subsurface, the air permeability of soil is significantly influenced by the density and viscosity (which are temperature dependent) of the soil gas. Soil gas with higher density and viscosity will be less mobile and less affected by the vacuum created by the SVE system. This will increase the number of extraction wells that are needed to effectively control the entire contaminant plume.

Even though air permeability tests can be conducted in the laboratory from soil samples collected from the field, it is a difficult task to extrapolate laboratory results to characterize the actual subsurface conditions of a site. The best way to characterize the air permeability of a specific site is to conduct field air permeability tests. Typical field air permeability test determination entails measuring pressure responses from air piezometers installed in the subsurface from a specific distance away from the SVE well.

The pressure versus time relationships can be expressed as follows (Johnson et al., 1990):

$$P = \frac{Q}{4\,\pi m(k/\mu)}\left[-0.5772 - \ln\left(\frac{r^2 \varepsilon \mu}{4\,kP_{atm}}\right) + \ln(t)\right] \tag{5.5}$$

where P = pressure measured at piezometer at a distance r and time t (g/cm-sec^2)

Q = volumetric vapor flow rate from extraction well (cm^3/sec)

m = stratum thickness (cm)
k = soil permeability to air flow (cm^2)
μ = viscosity of air (1.8×10^{-4} g/cm-sec)
r = radial distance from vapor extraction well (cm)
ε = soil void fraction
P_{atm} = atmospheric pressure (1 atm = 1.013×106 g/cm-sec^2)
t = time (sec)

Based on the above relationship, the field data are plotted on a semilog plot and the slope (S) of the best-fit regression line is determined. With the slope, soil permeability (k) can be calculated by:

$$P = \frac{Q}{4\,\pi m(k/\mu)} \quad \text{and} \quad k = \frac{Q\mu}{4\,S\pi m} \tag{5.6}$$

5.2.2.2 Soil Type

Coarse-textured and poorly graded soils will generally have higher permeabilities than fine-textured and well-graded soils. This is due to the larger pore spaces present in material such as gravels and sands and to the interconnections of the pore spaces.

5.2.2.3 Soil Moisture Content

High soil moisture content affects SVE negatively. The vapor extraction process is slowed down because moisture in the form of adsorbed moisture and/or moisture held in the capillary fringe interferes with the extraction process. In addition, vapor extraction in soils with high moisture content will result in high volumes of condensate collecting in the system.

5.2.3 Site Conditions

Site conditions refer to both above-ground and below-ground conditions. These can be either natural or man-made. Site conditions that can affect the effectiveness of an SVE system include the depth to water table, existence of preferential pathways, subsurface conduits, and surface caps.

5.2.3.1 Location of Water Table

An SVE system would not be effective in extracting contaminant vapors if the groundwater table is very high (5 ft below ground level, or less). High water levels will cause the SVE system to shut down frequently due to upwelling of groundwater and/or buildup of water in the condensate separator. One solution to this problem is to pump groundwater out of the SVE well while the SVE system is in operation. The simultaneous pumping of groundwater will not only eliminate upwelling problems but also create a cone of depression, thus exposing more areas for vapor extraction.

A high water table can short-circuit an SVE system. Instead of extracting soil vapor from the subsurface, the SVE system installed in areas with a high water table might pull air from the atmosphere through cracks on the ground surface or underground utility conduits. To rectify this problem, an impermeable surface cap will have to be installed around the SVE well. As a general rule of thumb, SVE alone will not be an effective remedial option if the groundwater level is less than 5 ft below ground level.

5.2.3.2 Subsurface Conduits

Typical subsurface conduits, such as electrical, telephone, sewer, and other utility lines, are bedded in sand, which is usually more permeable than the surrounding soils. As a result, the presence of subsurface conduits will usually create preferential flow pathways that reduce the effectiveness of an SVE system. Where practical, it is advisable to locate SVE wells away from subsurface conduits. The location of subsurface conduits will also provide a pathway of least resistance for vapors to migrate off-site. This can create potentially hazardous conditions outside of the site boundaries.

5.2.3.3 Surface Caps

A surface cap is an impermeable surface seal that prevents atmospheric air from entering the subsurface near the extraction well, which in turn short-circuits an SVE system. The presence of a surface cap can increase the radius of influence of the SVE system by forcing air to be drawn in peripherally from a greater distance. This will bring the flux into contact with a greater volume of soil, enhancing both the vaporization and extraction process.

Different materials can be used to construct a surface cap. Concrete and asphalt are the two most common materials that will serve as an effective cap. An alternative to concrete and asphalt is a clay or bentonite layer, which can be varied in thickness. Another alternative, synthetic polymer liner, has seen some application as surface cap construction material.

5.2.4 Design Considerations

If contaminant characteristics, soil properties, and site conditions are favorable, it is the engineer's role to design a system that will optimize the cleanup process in terms of cost, duration, and effectiveness. An overly conservative design will be costly to construct, maintain, and operate. Equipment such as blowers and vapor abatement units will have to be oversized to meet the requirements of an overdesigned system. However, because remediation

Table 5.3 Typical SVE Radius of Influence

Soil Type	Radius of Influence (ft)
Coarse Sand	>100
Fine Sand	60–100
Silt	20–40
Clay	<20

engineering deals with problems that are below ground and, thus, not visible, it possesses inherent difficulties insofar as enabling accurate designs.

It is, thus, the engineer's challenge to design a system that will adequately address the contamination problems based on best available data and estimates, while at the same time incorporating modifications into the design that will enable easy future expansions of the system, if necessary. Some of the design considerations that can affect the performance of an SVE system are briefly described in the following sections.

5.2.4.1 Radius of Influence

The radius of influence (ROI) is the distance at which the subsurface vacuum is approximately 0.1 inch of H_2O or 1 to 10% of the applied vacuum at the vapor extraction well. Typically, a pilot study will provide a better estimation of the ROI. Typical ROI that are observed in different types of soil formations are tabulated in Table 5.3.

5.2.4.2 Blower Size

Blowers are used to induce subsurface air flow necessary for soil vapor extraction by creating a negative pressure gradient in the surrounding soils to the SVE wells. The blower of an SVE system should be sized to effectively pull vapors from all of the vapor extraction wells. Three types of blowers that are commonly used for SVE applications are regenerative blowers, centrifugal blowers, and rotary lobe blowers.

Regenerative blowers are the most commonly used blowers for SVE systems. They can achieve high vapor flow rates with low to moderate vacuums (<120 in. of H_2O) and require minimal maintenance.

Centrifugal blowers can achieve very high flow rates, but can operate only at very low vacuums. They are appropriate only in sand and gravel environments or trench systems that are highly permeable in which the induced vacuum required to sustain high air flow rates need not be high. The advantages for using this type of blower include relatively low equipment cost, low electrical consumption, and low maintenance requirement.

Rotary lobe blowers are capable of producing very high vacuum (in the range of 10 to 20 in. Hg), but require frequent maintenance (monthly). They

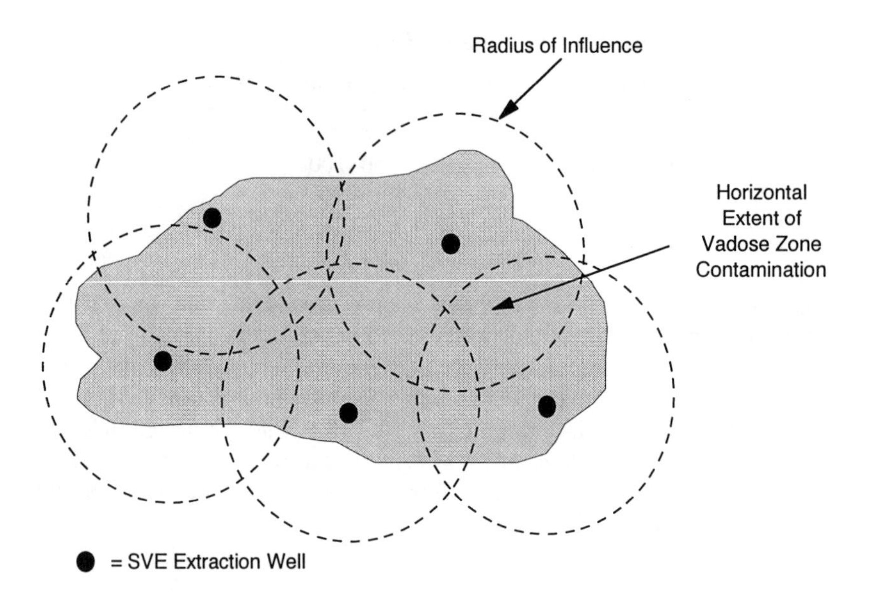

Figure 5.2. Extraction well locations and spacing.

are suitable for silt or clay environments that have low permeabilities. The disadvantages for this type of blower include relatively high equipment cost, high electrical consumption, high noise levels, and high maintenance requirement.

5.2.4.3 Extraction Well Locations and Spacing

Vapor extraction wells should be installed to cover the entire contaminated area to maximize the recovery of contaminants and to expedite the cleanup time. Ideally, the number of extraction wells required is based on the ROI of the SVE system using enough wells to overlap and encompass the area of soil contamination. Figure 5.2 illustrates typical extraction well locations and spacings.

However, more often than not, the overlapping of the ROI of the vapor extraction wells is not possible because of the presence of structures and other improvements. In these cases, vapor extraction wells are typically installed around the structure to cover the contaminated area as much as possible. Stub-outs should also be incorporated into the final design to account for future expansion of the entire SVE system. Underground pipings for the SVE system are laid in locations where future extraction wells may be required.

5.2.4.4 Extraction Well Screens

Placement of screen intervals of extraction wells should be a carefully thought-out process. Figure 5.3 is an isobar and flownet diagram illustrating

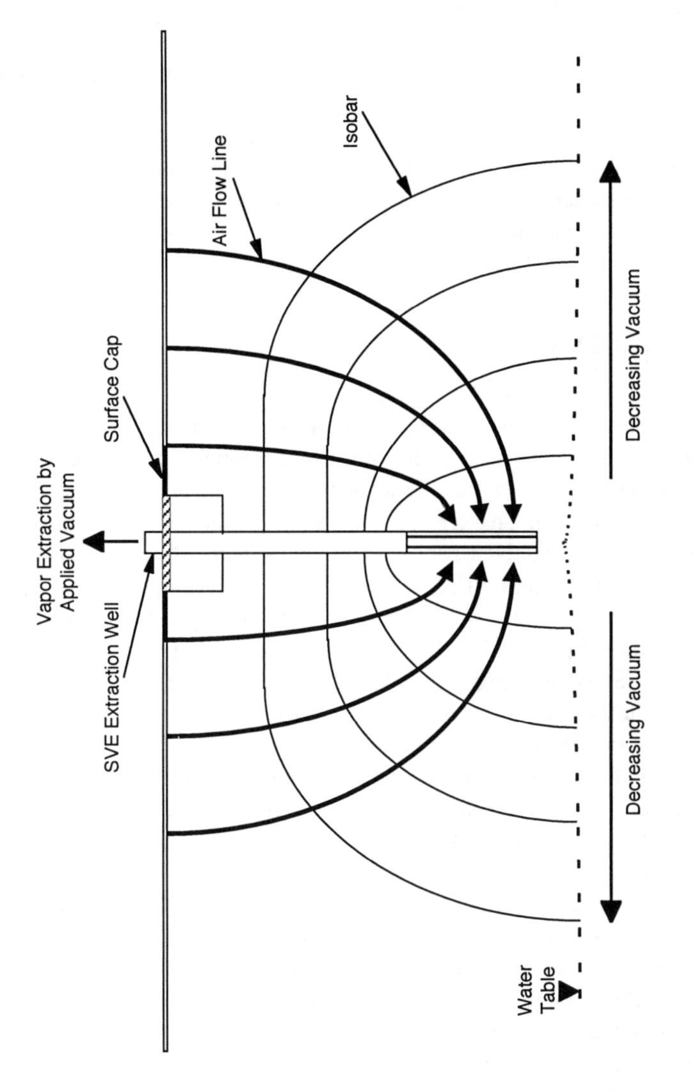

Figure 5.3. Relationship between induced vacuum and air flow around extraction well.

the relationship between induced vacuum and air flow around an extraction well. It is generally accepted as a rule of thumb by many in the industry that the optimum screen length for an extraction well should be approximately 10 ft; 15 ft is considered the maximum. The reason for this is that beyond 15 ft, the applied vacuum would have dissipated to a negligible level and would have been rendered ineffective in inducing any air flow.

However, in a pilot test conducted by the authors in homogeneous sandy soil, a radius of influence of up to 140 ft was observed for an extraction well screened between 60 and 110 ft (HWS Consulting Group, 1995). The observation well had a 6-in.-length screen at about 90 ft deep and a detectable flux of 1.63 L/(min-ft). This clearly indicates that, at least for homogeneous sandy soils, significant flux can be encountered beneath the 15-ft maximum guideline. Extraction wells are typically 2 to 4 in. in diameter. Doubling the well size will only increase the air flow rate by about 15%.

5.3 AIR SPARGING CONCEPT

Like SVE, air sparging (AS) operates on the principle that petroleum hydrocarbons can volatize at temperatures typically found in the subsurface. Unlike SVE, AS introduces air into the subsurface through direct injection of compressed air as opposed to creating a vacuum and drawing in air from the perimeter. As described in Chapter 3, contaminants in the saturated zone can exist in three phases: dissolved phase (in groundwater), adsorbed phase (to the soil particles), and liquid phase, if free product is present.

Air sparging is basically the injection of air under pressure below the water table for the purpose of volatilizing the different phases of contaminants in the groundwater to the vapor phase. This flow of air creates a transient air-filled porosity in the aquifer or water-bearing zone by the displacement of water in the soil matrix. The minimum pressure required to displace water in an AS system is that which is needed to overcome the resistance of the soil matrix to air flow. This resistance to flow is a function of the height of the water column that needs to be displaced and the flow restriction (air/water permeability) of the soil matrix. Additionally, plugging of the well may increase the resistance to air flow and add to the head loss. When this "breakthrough" pressure is achieved, air enters the soil matrix displacing water, and eventually exits into the vadose zone. First approximation of the breakthrough pressure is usually a number slightly greater than the hydrostatic head.

Air sparging must operate together with SVE. As described in the previous section, SVE systems basically remediate contaminated soils through two actions: mass transfer (dissolved and liquid phases volatilizing to vapor phase) and mass transport (extraction). Air sparging provides both mechanisms too, mass transfer and mass transport (from the saturated zone to the vadose zone); however, the mass transport stops at the vadose zone. Without the vapor extraction mechanism provided by the SVE system, AS would not work. In fact, AS could potentially increase both laterally and vertically the extent of contamination migration as it would create a net-positive subsurface pressure. Besides expediting the expansion of the contaminant plume, the uncontrolled migration of contaminated vapors could create both health and explosion hazards.

In the mass transfer process, both AS and SVE share the same fundamental concepts. However, the mechanism or nature of mass transport in AS is not as straight-forward as in the SVE process. Two schools of thought are available to explain the phenomenon (U.S. EPA, 1990a). The first theory suggests that sparging creates air bubbles that move through the groundwater to the vadose zone. The second suggests that the sparged air moves through pathways in the saturated zone to the vadose zone. In both cases, the sparged air moves upwards in the aquifer due to the difference in density between air and water. As with SVE, the same groups of criteria affect the utilization of AS as a remedial option. These conditions are discussed below.

5.3.1 Contaminant Characteristics

Since both AS and SVE operate on the same principles of contaminant volatilization, the same contaminant characteristics that apply to SVE will apply to AS.

5.3.2 Soil Properties

The soil properties that are critical for the success of SVE are just as important for AS, except for soil moisture content, which technically would not be a consideration because sparging is a saturated zone process. However, since AS and SVE operate in tandem, it follows that soil moisture content will be a factor in the overall treatment process. The most important considerations in soil properties for air sparging are permeability and soil types.

5.3.2.1 Permeability

In air sparging, soil permeability is important in two respects: (1) the vertical permeability has to be high enough to allow air to rise through the saturated zone into the vadose zone, and (2) the horizontal permeability has to be high enough to allow a fair amount of lateral movement of the injected air. In glacial till or fractured consolidated deposits, a significant portion of the injected air may flow through fractures or channels, thus reducing the effectiveness of AS.

5.3.2.2 Soil Types

As for SVE, AS works best in coarse-textured and poorly graded soils as opposed to fine-textured and well-graded soils. This is due to the larger pore spaces present in material such as gravels and sands, and to the interconnections of the pore spaces.

5.3.3 Site Conditions

Site conditions that can affect the effectiveness or viability of an AS system include depth to the water table, existence of preferential pathways, subsurface conduits, confining layers, and the presence of free-floating product.

5.3.3.1 Location of Water Table

Location of the water table under all seasonal conditions should be estimated for optimal AS system operation. This is because the top of the screen of the AS wells should be set beneath the contaminant plume for the AS system to be effective. If the seasonal low static water tables are not properly estimated, the AS system will be of no use when the water table drops beneath the well screen during the dry season.

In addition, all of the AS wells should be installed with the top of the well screens set at approximately the same depth from the water table. Typically, a number of AS wells will be manifolded together and all wells operated at an equal pressure. If the top of the well screen of one well is installed much closer to the water table than the others in the system, most of the air will pass through this well because less pressure is required to depress the shorter water column above the well screen. This will significantly reduce the effectiveness of the overall AS system.

5.3.3.2 Confining Layers

The presence of confining layers, such as clay lenses, will adversely affect the efficiency of an AS system because the vertical migration of air to the water table surface will be severely limited. At locations where the site geology is characterized by intermittently mixed layers of materials with varying degrees of permeability, sparging should be approached with caution. There is a twofold danger in injecting pressurized air into a saturated zone with an overlying zone of lower permeability: (1) the sparged air and volatilized hydrocarbon vapors will likely migrate off-site along pathways beneath the overlying cap, and (2) vapor extraction wells (which are screened above or in the confining layer) will not be able to remove the sparged vapors.

5.3.3.3 Presence of Free-Floating Product

Air sparging systems should not be implemented at a site with free-floating product because the upwelling effect created by the sparging at the sparge wells may cause the free-floating product to migrate to previously uncontaminated areas. At a site with free-floating product, groundwater pumping and product recovery should be installed first to recover as much free-floating product as possible before an AS system can be implemented.

5.3.3.4 Subsurface Conduits

Subsurface conduits and other man-made avenues of preferential pathways are always a problem when volatilization and flux creation are involved. However, because subsurface conduits are typically not installed at depths beneath the water table, they should not be a problem for air sparging.

5.3.4 Design Considerations

As with SVE design considerations, the engineer can optimize the design to achieve a system that is capable of remediating the site without running up construction costs. The components of an air sparging system include the blower or air compressor, sparge wells, piping and manifold, and pressure and flowrate controls. Design considerations will take into account injection pressures, injection flowrates, sparge well spacings, and sparge well screen depths.

5.3.4.1 Radius of Influence

The combined ROI of the AS systems should be superimposed on the combined ROI of the SVE system. The combined AS ROI should be located within the combined SVE ROI, with a safety margin of at least several feet.

5.3.4.2 Injection Pressure and Blower/Compressor Sizing

The injection of air under pressure below the water table in an AS system is accomplished by either a blower or an air compressor. Blowers are cheaper than air compressors; however, they are limited by the amount of pressure they can provide. Most blowers are capable of providing a maximum pressure of 15 psi. Beyond that, an air compressor will be required. Depending on the individual state regulations, an oil-less air compressor may have to be specified. Oil-less air compressors cost more than regular air compressors.

The minimum pressure that is required to displace water in an AS system is basically equal to the pressure that is needed to overcome the overall resistance of the soil formation in the saturated zone to air flow. This resistance is a direct function of two important factors: (1) the height of the water column in the sparge well that needs to be displaced, and (2) the flow resistance of the soil matrix itself that is a function of the air/water permeability of the soil type. Additionally, a badly plugged well may increase restrictions to air flow and add to the head loss.

In calculating the breakthrough pressure that is required, the hydrostatic head (water column in the well above the screen) is a good first assumption. Unless highly permeable soil formations are encountered, the breakthrough pressure will usually be some multiple of the hydrostatic head. The hydrostatic head is determined by measuring from the water table down to the top of the slotted section of the sparge well. Depending on the type of soil, additional pressure will be required to overcome capillary effects due to capillary rise within the aquifer. This value is generally ignored when determining the hydrostatic head.

The air pressure required to sustain a given air flow into the aquifer is usually referred to as the injection pressure. Typically, to maintain design flowrates, injection pressures at 1.25 to 2.0 times the hydrostatic head are applied. The authors have also encountered numerous situations where the injection pressure is equal to the hydrostatic head. Care should be exercised when regulating sparge pressures. Excessively high pressures can cause fracturing of the soil. This is preceded by an increase in pressure, followed by a sudden drop in pressure as new channels are created by the fracture.

5.3.4.3 Sparge Well Locations

The location of the sparge well is probably the single most important consideration in the design of a SVE/AS system. Sparge wells cannot be located where SVE wells may not be able to capture the sparged hydrocarbon vapors. There should always be a cluster of SVE wells in which the combined ROI will encompass the ROI of the sparge wells in that area. Because of the potential dangers of creating pathways to sensitive receptors, the rule of thumb is that sparge wells should be conservatively rather than ambitiously located.

5.3.4.4 Sparge Well Screens

The screened interval or slotted section of the sparge well should be located at the bottom of the contamination plume in the water-bearing zone or aquifer of concern. However, at most typical service station and similar-size sites, the vertical distribution of contaminant concentrations in the saturated zone is usually not contoured. Depending on the thickness of the saturated zone, the slotted section is generally located close to the bottom of the water-bearing zone or as deep as possible, but it is not usually deeper than 20 ft beneath the water table. The screen interval is usually between 1 to 2 ft in length.

5.4 PILOT TESTS

Pilot tests are almost always conducted prior to the final design and installation of SVE and SVE/AS systems. The tests are relatively simple and straight-forward to perform and provide useful information on the feasibility of a site to SVE and SVE/AS treatment. Additionally, the data are essential for designing the final system.

Two issues are oftentimes raised when carrying out SVE/AS pilot tests; the first pertains to the length of the tests, and the second to the number of extraction wells tested. There are no clear answers to either question. Budgetary constraints probably play as big a role as any other reason in determining the length and level of detail and effort with which a test is conducted. From an engineering standpoint, an SVE/AS test should be performed until steady-state conditions have been achieved. Strictly speaking, design data collected under steady-state conditions is the only valid data to be used in the final design process.

Field experience has demonstrated that the length of time required to achieve steady-state conditions is a function of site conditions and local geology. In homogeneous sandy soils, steady-state conditions can be achieved within the hour; however, in complex geological formations with interbedded layers and lenses of materials with varying permeabilities, it might be days before a steady set of data is observed. Similarly, a site with complex geology and more than one distinct vertical zone of contamination may require tests performed on several wells with different screen intervals.

The following sections will describe the general protocol that is typically followed in conducting SVE, AS, and SVE/AS tests.

5.4.1 SVE Pilot Test

5.4.1.1 Purpose

The purpose of the SVE pilot test is to determine the feasibility of soil vapor extraction as a method of soil contamination cleanup. Data and information collected during the test are then used to:

1. Size the blower (what horsepower).
2. Determine the radius of influence (how far apart to place the extraction wells).
3. Determine vapor treatment requirements.
4. Estimate length and cost of cleanup operation.
5. Evaluate other design parameters.

To achieve item (1), the approximate operating applied vacuum (pressure on suction side of the blower) and the total system air flowrate (air flowrate per extraction well multiplied by the number of extraction wells) need to be determined. By plotting a graph of applied vacuum versus air flowrate, the optimum vacuum and its corresponding flowrate per extraction well can be determined. As such, the pilot test should be conducted at several vacuums.

To achieve item (2), information regarding attainable air flow rates with increasing distance from the extraction well needs to be gathered. By plotting a graph of distance from extraction well and corresponding air flowrates, an optimal and cost-effective radius of influence can be decided. As such, air flow-rates induced at vapor monitoring wells of varying distances need to be measured.

To achieve items (3) and (4), contaminant vapor concentrations need to be measured. Item (5), other parameters, include measuring depths to water in the extraction well during the test to determine the effects of upwelling.

5.4.1.2 Set-Up

The location of the extraction wells should be within the plume of high contaminant concentration. Site conditions and restrictions will certainly play an important role in the selection process. The wells will most likely be incorporated into the final SVE design. Well screens should correspond to depths where contamination was identified on the boring logs.

Monitoring well distances from the extraction well are determined based on site geology, experience with similar sites, and also site restrictions. Avoid placing the monitoring wells in close proximity of existing wells as this might create a short-circuiting effect. Figures 5.4 and 5.5 illustrate the construction details of typical extraction and monitoring wells. The blower and other appurtenances, including the vapor treatment unit, can be conveniently mounted on a trailer and moved from site to site.

5.4.1.3 Equipments and Measurement Instruments

The extraction unit typically consists of a regenerative blower equipped with an explosion-proof motor enclosure. In some states, the effluent from the blower has to be routed to an air-treatment unit. The off-gas treatment methods usually employed are internal combustion engines (ICE), oxidizers, and activated carbon. The use of carbon may prove to be expensive, as the concentrations during pilot tests are usually high. Internal combustion engines are capable of generating high vacuums; however, the extracted soil vapor must contain a minimal level of oxygen to maintain combustion. Usually, when the hydrocarbon concentrations are high, the corresponding oxygen levels will be low. Dilution air is then bled into the manifold to keep the air-to-hydrocarbon ratio in stoichiometric proportion. This presents the biggest drawback of an ICE because vacuum applied to the extraction well is compromised by the drawing in of dilution air. Many equipment vendors provide trailer-mounted complete extraction-treatment packages. The process piping should be configured and equipped such that data collection requirements can be met.

Variables that need to be measured are

1. Vacuum (negative pressure) applied at the extraction well(s) and induced at the monitoring wells.
2. Air flowrates induced at the extraction well(s) and the monitoring wells.
3. Contaminant concentrations.
4. Water level changes in the extraction well(s) and surrounding monitoring wells.

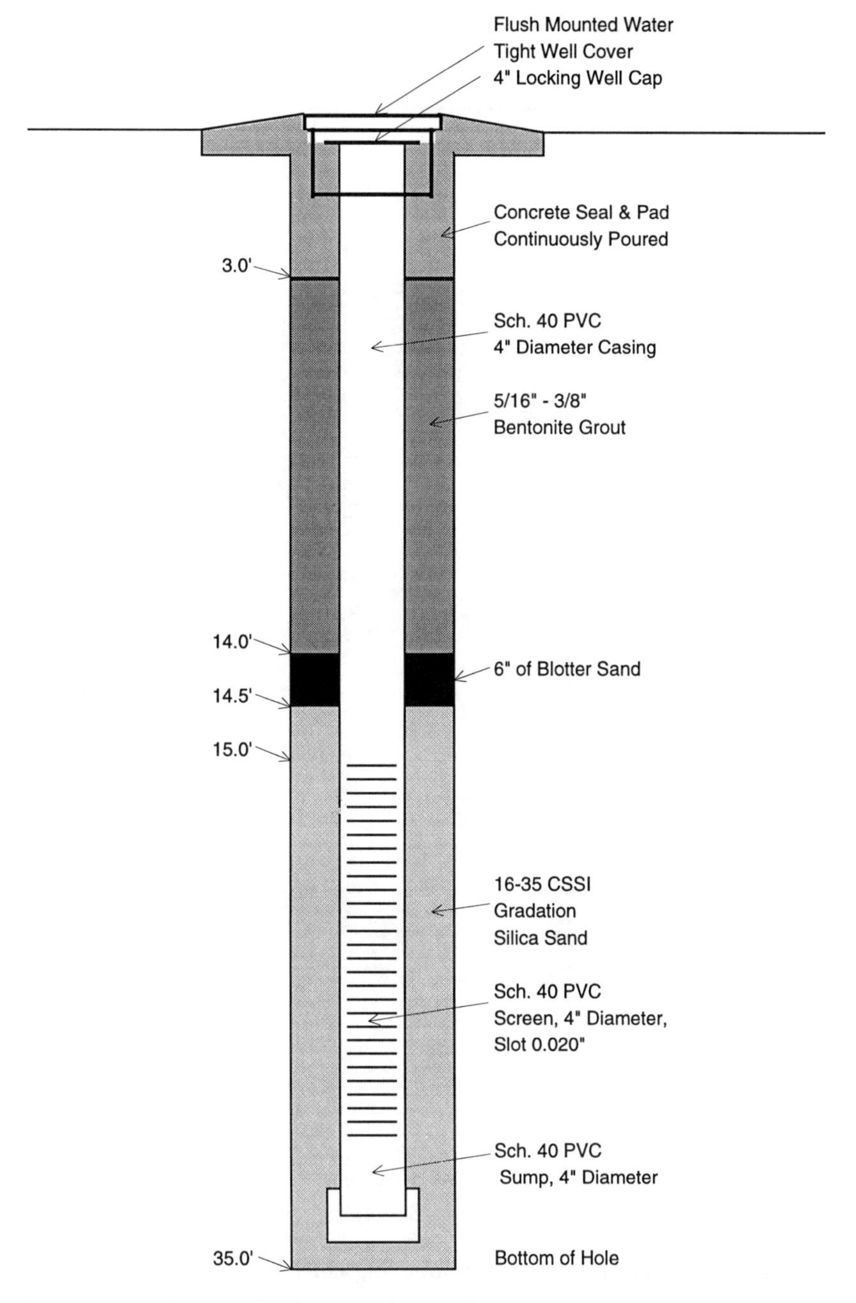

Figure 5.4. Typical SVE extraction well.

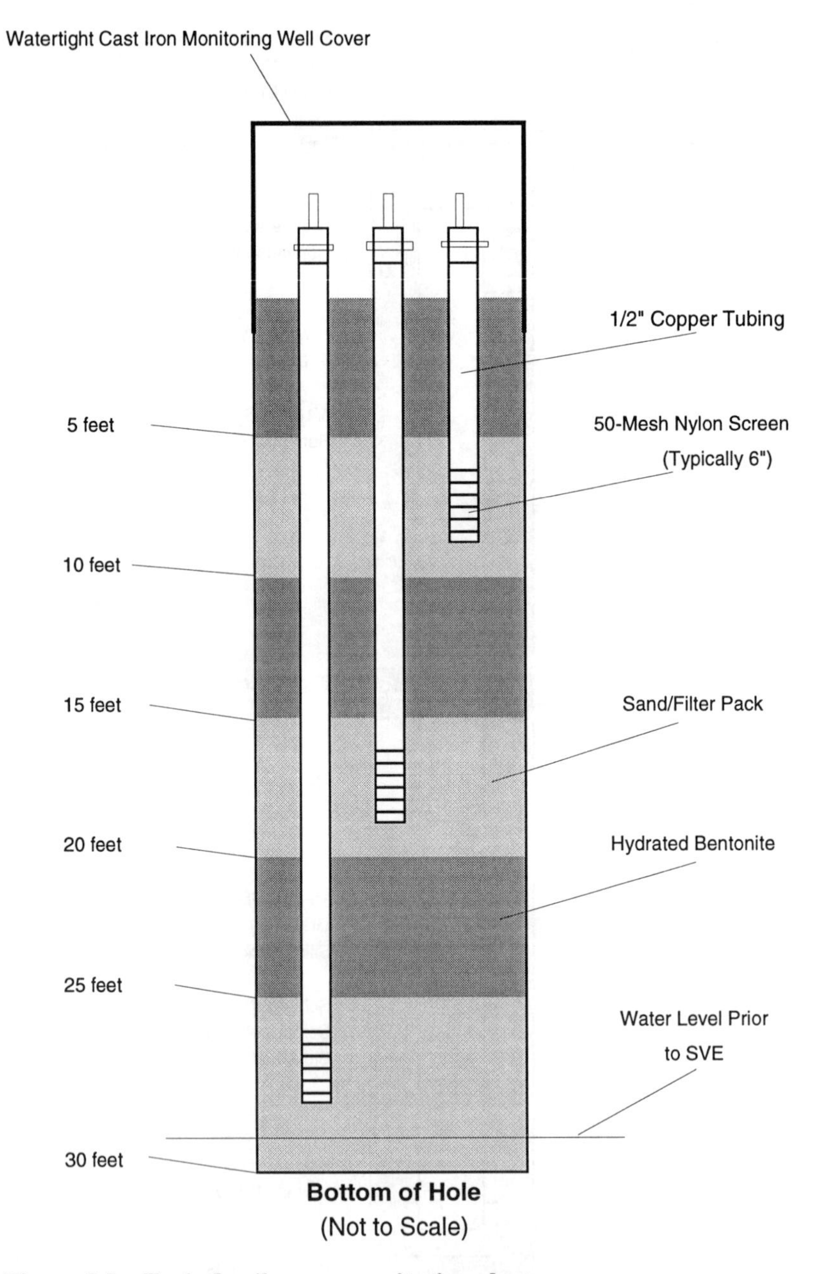

Figure 5.5. Typical soil vapor monitoring cluster.

Vacuum readings are measured with pressure gauges. Air flowrates at the extraction well are usually computed by measuring the velocity pressure. Velocity pressure (differential pressure) is typically measured with a pitot tube and pressure gauge. Velocity pressure is the difference between total pressure and static pressure (in this case, the vacuum applied). The air velocity is then obtained from manufacturer-provided air velocity versus velocity pressure graphs. Some pressure gauges have dual scales; one for velocity pressure and one for velocity.

Flowrate is obtained by multiplying velocity with the cross-sectional area of the piping. Velocity pressure can also be measured using a pitot tube and manometer. To obtain actual air flowrates (usually expressed in actual cubic feet per minute (ACFM)), corrections for barometric pressure, temperature, and relative humidity must be made. Manufacturer-provided formulae for specific instruments allow for the conversion from standard flowrates (observed flowrates) to actual flowrates. Some customized vapor extraction pilot test trailers are equipped with vacuum and flow read-outs in their control panel. However, these data should be verified somewhat with the above described measurements.

Air flowrates at monitoring wells can be measured using rotameters. Some caution should be exercised in the interpretation of these values as the rotameter is actually measuring air being sucked into the monitoring well from the atmosphere (the tubing from the monitoring well is connected to the top of the rotameter). However, as this movement of air eventually encounters the same resistance in the subsurface (soil air permeability) as the rest of the flux, the flowrates measured should provide a good idea of the magnitude of air flowrate induced.

In general, flow and velocity measuring instruments are calibrated to standard conditions. Standard conditions are defined as pressure at 29.92" Hg (1 atmosphere or 14.7 psi) and temperature at 70°F (some manufacturers use 68°F or 60°F). Despite the actual barometric pressure and temperature conditions, the instruments measure values based on standard conditions, therefore observed readings are referred to as standard values. Thus, the unit for observed flowrates is standard cubic feet per minute (SCFM). Corrected flowrates are presented in ACFM. The conversion from SCFM to ACFM is

$$\text{ACFM} = \text{SCFM} \times \frac{14.7}{(14.7 + \text{actual pressure})} \times \frac{(460 + \text{actual temperature})}{(460 + \text{standard pressure})} \tag{5.7}$$

5.4.1.4 Data Collection

A workplan is usually prepared prior to conducting the pilot test. The primary purpose of the workplan is to provide field personnel with a written outline on the procedures and protocol that need to be followed. As discussed earlier, decisions will have to be made on the number of wells to perform tests on, and the length of each test. In addition, monitoring wells will have to be selected. The engineer will also have to anticipate the different vacuums under which tests should be conducted. Site geology and experience will be the dictating factors. Sometimes, adjustments or deviations from the workplan may need to be made on-site.

In general, the pilot test at the selected extraction test well(s) will consist of several short-term tests and a long-term test. The short-term tests will be conducted at at least three different applied vacuums. The reason for applying different vacuums to the extraction well is to establish a relationship between vacuum applied and flow extracted. The following parameters will be monitored during each short-term test (along with examples of typical measurement instruments):

- Induced air flowrate from the extraction wells using a pitot tube velocity meter installed in the PVC pipe connecting the well to the blower or ICE
- Applied vacuum at the extraction wells and induced vacuum at the monitoring wells using magnehelic pressure gauges placed at each well to evaluate the ROI for the well at each applied vacuum
- Depths to water in the extraction well at the beginning and towards the end of each test

After evaluating the well air flow characteristics (from graph of applied vacuum versus induced air flowrate), the extraction well(s) will be subjected to the selected applied vacuum for the long-term test. The selected vacuum should be the optimum vacuum for the site, beyond which one would see a diminishing rate of return (airflow) for an increase in cost (applied vacuum). Data obtained from the monitoring wells will help determine the ROI of the extraction well(s) at the applied vacuum. The following parameters will be monitored during each long-term test:

- Induced air flowrate from the extraction well(s) using a pitot tube velocity meter installed in the PVC pipe connecting the well to the blower or ICE
- Applied vacuum at the extraction well(s) and induced vacuum at the monitoring wells using magnehelic pressure gauges placed at each well to evaluate the ROI for the well at each applied vacuum
- Percent oxygen in extracted vapor using an oxygen meter, and organic vapor concentrations in extracted vapors at extraction well(s) using a suitable volatile-hydrocarbon meter
- Lead analysis, if necessary
- Total petroleum hydrocarbons as gasoline (TPHg); BTEX; fixed gases (carbon dioxide, oxygen, and methane) in extracted vapor by

an air sample taken at the extraction well using Tedlar bags and a vacuum box

- Depths to water in the extraction well at the beginning and towards the end of each test

Table 5.4 is an example of a typical field data sheet used for recording SVE pilot test data.

5.4.1.5 Data Evaluation

1. Calculate the air flowrate in the extraction well from

$$Q = vA \tag{5.8}$$

where Q = flowrate, SCFM
v = velocity (from velocity meter)
A = cross-sectional area of pipe

2. Tabulate applied vacuum and air flowrates induced at the extraction well. For both the static and differential pressure values, use the average readings for a particular applied vacuum. Example of data generated is shown in Table 5.5.
3. Plot graph of applied vacuum versus air flowrate (Figure 5.6).
4. Select desired operating applied vacuum. Use 90 in. water column (WC) in the example because beyond this point we are entering the region of diminishing return (the point at which the relationship is no longer a straight line).
5. From data collected, calculate the detectable air flowrates (ACFM/ft of screen) at each of the vapor monitoring wells. The detectable air flowrate is obtained by dividing the flowrate by the length of screen intervals.
6. Tabulate the detectable air flowrates at the different applied vacuum with the distance of the monitoring wells from the extraction well as shown in Table 5.6.
7. Plot graph of detectable air flowrates at the monitoring wells versus their distance from the extraction well (see Figure 5.7). Select desired operating ROI.
8. With ROI determined and plume area already identified, the number and location of extraction wells can be determined.
9. Determine total system air flow, which is the air flow per extraction well multiplied by the number of extraction wells.
10. With known operating applied vacuum and the total system air flowrate, the blower can now be sized from the manufacturer's blower graphs.

Table 5.4 Typical Field Data Sheet for SVE Pilot Test

Soil Vapor Extraction Test Field Data

Site: ____________________
Location: ____________________
Job: ____________________
Site Personnel: ____________________

Date	Time	Hours	Extraction Well			Monitoring Wells (Distance from SVE Well in Feet)					
						MW-1		MW-2		MW-3 (75)	
			Applied Vacuum (in. WC)	Flowrate (cfm)	DTW (feet)	Vacuum (in.WC)	Flowrate (cfm)	Vacuum (in.WC)	Flowrate (cfm)	Vacuum (in. WC)	Flowrate (cfm)

Table 5.5 Air Flow at Different Applied Vacuum

Applied Vacuum (in. WC)	Calculated Air Flowrate (ACFM)
45	27
60	36
75	44
90	50
120	53

11. During the pilot test, contaminant concentrations will be taken at both the suction (influent) side and pressure (effluent) side at certain intervals. Theoretically, at least three readings are needed: initial, after one pore volume has been extracted, and at end of test. Both field readings and laboratory samples are collected. Field measurement devices include sensidyne tubes and PID.
12. During the pilot test it will be necessary to estimate the pore volume value. This can only be calculated after the ROI is somewhat reliably known. The example below illustrates the determination of pore volume.

Example: To determine one pore volume.

Parameters needed:
Radius of influence = 50'
Length of extraction well screen = 30'
Soil density = 150 lb/ft^3
Soil porosity = 30%
Water content (moisture content) = 4%
Flowrate = 120 ACFM

$$\text{Water Content} = \frac{\text{Weight of Water } (W_w)}{\text{Weight of Solids } (W_s)}$$

$$0.04 = W_w / W_s \tag{5.9}$$

$$W_s = 25\ W_w$$

$$\text{Soil Density} = \frac{W_s + W_w + W_{air}}{V_{total}}$$

$$= \frac{(W_s + W_w)}{V_{total}} \quad \text{because } W_{air} \text{ is negligible} \tag{5.10}$$

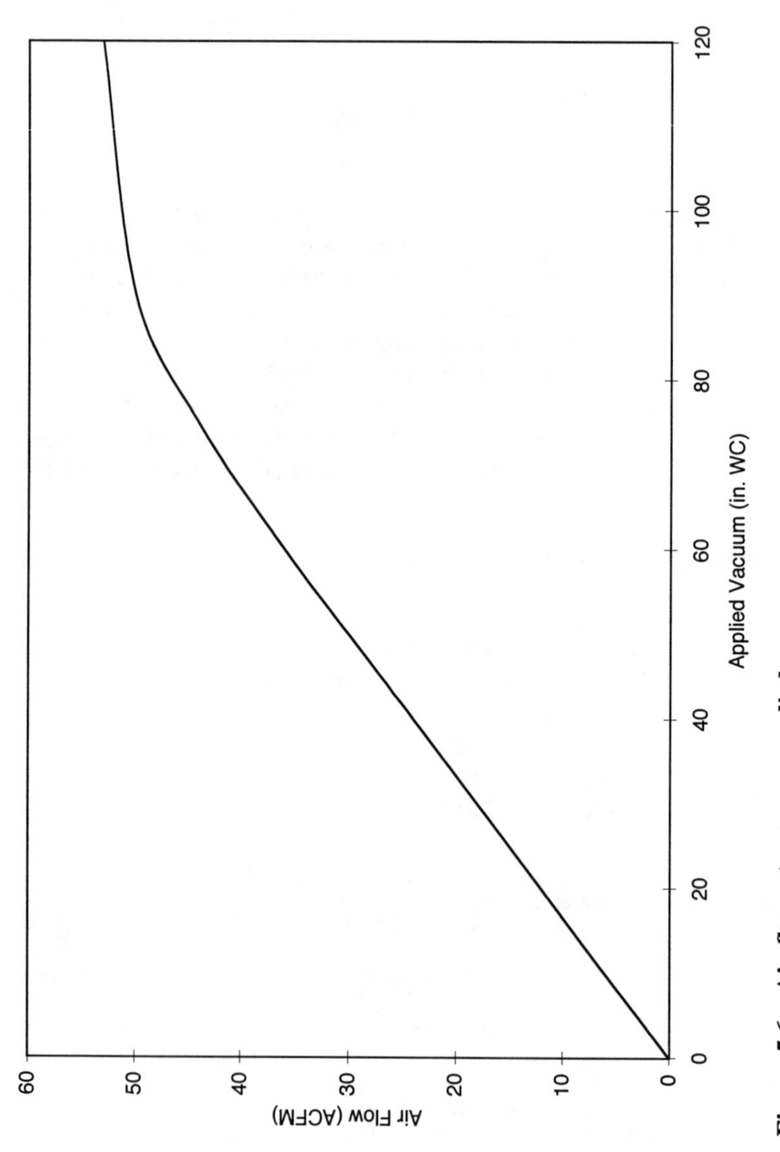

Figure 5.6. Air flowrates versus applied vacuum.

Table 5.6 Detectable Air Flow vs. Radial Distance

Vacuum Applied at Extraction Well (in. WC)	Air Flow at Monitoring Wells (ACFM)		
	MW-1	MW-2	MW-3
	Distance from Extraction Well (ft)		
	20	45	75
45	0.159	0.070	0.025
60	0.177	0.084	0.034
75	0.193	0.091	0.043
90	0.219	0.100	0.052
120	0.265	0.124	0.070

In one cubic foot of soil,

$$150 \text{ lbs} = W_w + 25\ W_w$$

$$= 26\ W_w$$

$$W_w = 150/26$$

$$= 5.77 \text{ lbs}$$

Volume of water in one cubic foot of soil

$$= 5.76 \text{ lbs}/62.4 \text{ lbs/ft}^3$$

$$= 0.09 \text{ ft}^3$$

$$V_{air} = V_{void} - V_{water}$$

$$= 0.3 - 0.09$$

$$= 0.21\ \text{ft}^3/\text{ft}^3 \text{ of soil}$$

$$\text{Percent } V_{air} = 0.21/1.0$$

$$= 0.21$$

$$= 21\%$$

$$\text{One pore volume of air} = \left(\pi r^2 \times \text{depth}\right)\text{ft}^3 \times \text{percent } V_{air}$$

$$= \left(\pi \times 50^2 \times 30\right) \times 0.21$$

$$= 49,486.5 \text{ ft}^3$$

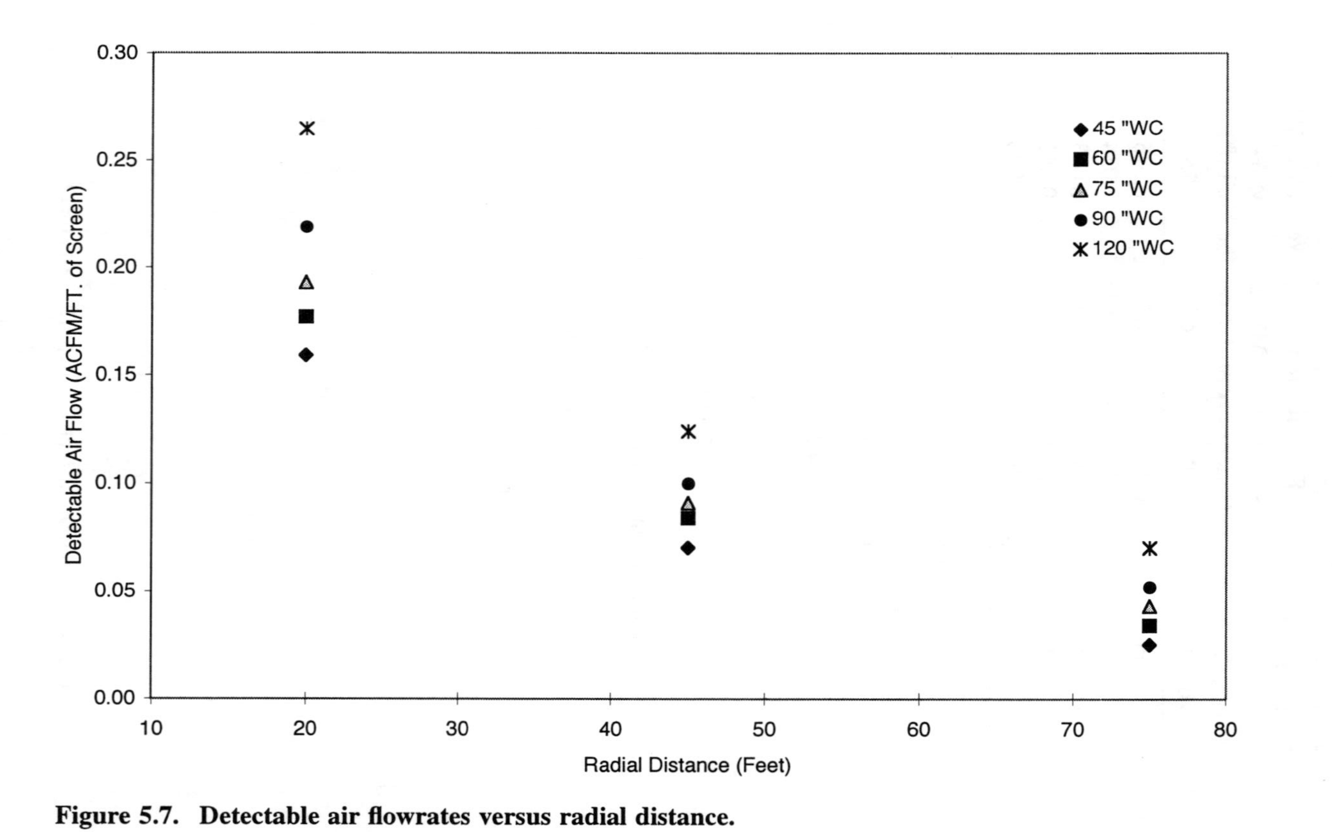

Figure 5.7. Detectable air flowrates versus radial distance.

$$\begin{aligned}\text{Extraction time} &= \text{one pore volume of air/flowrate}\\ &= \left(49{,}486.5\ \text{ft}^3\right)/\left(120\ \text{ft}^3/\text{min}\right)\\ &= 412\ \text{min}\\ &= 6.9\ \text{hours}\end{aligned}$$

5.4.2 Air Sparge Pilot Test

5.4.2.1 Purpose

Air sparging pilot tests are usually conducted in two phases. The first phase involves running the sparge test alone, and the second phase involves running both the sparge and SVE tests concurrently.

The purpose of conducting the AS test is to evaluate the feasibility of air sparging as a remedial technology and to determine AS design parameters. Data collected during the test can be used to:

1. Determine the radius of influence.
2. Determine airflow pathways.
3. Size the air compressor/blower.
4. Evaluate effects of mounding.
5. Evaluate effectiveness of sparging.

To achieve items (1) and (2), helium, dissolved oxygen (DO), and VOC concentrations in the monitoring wells need to be measured. Measuring applied pressures and air flowrates in the sparge well enables determination of pressures required. Taking depth to water level readings in the monitoring wells provides information on the effects of mounding. This is important as water mounding can block off screen intervals of wells, thus reducing or eliminating flux from entering the extraction wells during SVE. Monitoring the changes in VOC concentrations in the monitoring wells will help evaluate the effectiveness of sparging.

5.4.2.2 Set-Up

Typically, the well screen or slotted section of the air sparge well is approximately 1 to 2 feet in length. Theoretically, the well screen should be located near the vertical nondetectable line of the plume or near the bottom of the water-bearing zone of concern. However, this is seldom practical. For light non-aqueous phase liquid (LNAPL) plumes, screen depths of 10 to 20 ft below the water table are typically used.

Helium gas is used as a conservative tracer gas to determine airflow pathways in the subsurface. Mixtures of 20% to 60% of helium gas in air are fairly common.

5.4.2.3 Equipment and Measurement Instruments

The air sparge test unit will consist of compressed air and compressed helium supplies. Each of the gas flowlines should be equipped with appropriate pressure regulators, throttle valves, pressure indicators, flowrate indicators, and temperature indicators. The compressed air source should be equipped with a coalescing oil removal filter rated at 0.01 microns.

Wellhead assemblies for the monitoring wells should facilitate collection of sparged gas samples from both the vadose and saturated zones. Tubings for collection of saturated zone sparged gas must be of sufficient length to penetrate the water table. The wellhead assemblies should also enable measurement of depths to water levels. Field instruments for on-site measurements of DO, VOC, and helium should be available; FID, PID, or LEL meter can be used for measuring VOC concentrations, whichever is more sensitive.

5.4.2.4 Data Collection (Gasoline-Contaminated Site)

1. Measure and record depths to groundwater in the sparge well and monitoring wells.
2. Collect groundwater samples from the sparge well and monitoring wells for the following analyses:

Field:	dissolved oxygen (DO)
Laboratory:	BTEX (optional)
	TPHg (optional)

3. Install the appropriate wellhead assemblies onto the monitoring wells that will facilitate the collection of gas samples from both the vadose and saturated zones.
4. Collect vadose zone gas samples from the monitoring wells for the following analyses:

Field measurement:	VOC
	helium
Laboratory analyses:	BTEX (optional)
	TPHg (optional)

5. Collect saturated zone gas samples from the monitoring wells for the following analyses:

Field measurement:	helium

6. Install sparge wellhead assembly onto the sparge well.

7. Calculate head of water above bottom of screen from depth to water measurements. For example, if the top of the screen is 15.0 ft below the water table, then the hydrostatic head to be overcome would be 15.0/2.31 = 6.5 psi.
8. Deliver compressed air. First approximations of the breakthrough pressure is usually some number slightly greater than the hydrostatic head, in this case assume it to be 120% of the water column, approximately 8.0 psi.
9. Gradually increase air pressure to the 6.5 to 8.0 psi range. Continue increasing the pressure until breakthrough is reached. Record the breakthrough pressure.
10. Increase the pressure until the lowest desirable flowrate is achieved. Perform the first test at this pressure and flowrate.
11. Deliver helium gas at the same pressure as the compressed air.
12. During the test, collect the following data and samples:

    Air and helium flowrates and pressure at the sparge well.
    Depth to water in monitoring wells (sparge well will be dry).
    Soil gas samples from monitoring wells.
    Field measurement: helium — vadose zone and saturated zone
    VOC — vadose zone
    Pressure in monitoring wells.

13. Stop test. Perform the following:

    Depth to water in all wells.
    Collect vadose zone soil gas samples from monitoring wells for the following analyses:
    Laboratory analyses: BTEX (optional)
    TPHg (optional)
    Collect water sample from all wells for the following analyses:
    Field measurement: DO

14. Repeat the test at two higher applied pressures.
15. Table 5.7 is an example of a typical form used for the collection of data during an AS test.

5.4.2.5 Data Evaluation

Based on the data collected, three primary parameters need to be defined:

1. AS radius of influence.
2. AS pressures and flowrates required.
3. Existence of preferential pathways.

The AS ROI is estimated based on observations of increases in helium, VOC, and DO concentrations in observation wells. Measurement of induced positive pressures at observation wells will also provide an idea of the AS radius of influence. It should be kept in mind that the ROI based on pressure observations is theoretically greater than the actual ROI of the AS because the pressure response is transmitted farther than the radius of air bubble migration through the aquifer to the water table. However, where soil conditions dictate that lateral permeability is significantly greater than vertical permeability, the reverse could be true.

An optimum pressure and, thus, flowrate is selected based on the ROI generated. By analyzing the distribution of concentrations of helium in the observation wells, the existence of preferential airflow pathways can be established.

5.4.3 Combined SVE and Air Sparge Pilot Test

5.4.3.1 Purpose

A combined SVE and AS test is usually conducted after individual SVE and AS tests have been carried out. The combined air sparge–SVE test provides useful data for designing the overall system. The purpose of the test is to evaluate how the effects of simultaneously inducing a vacuum in the vadose zone and applying a positive pressure in the saturated zone affects the following:

1. SVE radius of influence.
2. Water level changes.
3. VOC concentration changes.

Theoretically, the SVE ROI should decrease as sparge flow increases. However, this is not always the case. If there is a sufficient decrease in the SVE ROI, the test should continue with various combinations of applied pressures and vacuums. Because SVE creates a mounding effect around the extraction wells while AS causes a mounding effect away from the sparge well, their combined effect on the available screen intervals of the extraction wells could be significant. Water level changes should be closely monitored during the test as this would assist in determining the location of SVE and AS wells in relation to one another. With AS, VOC concentrations should increase. This increase is crucial to the selection and design of abatement equipment.

5.4.3.2 Set-Up

The set-up for the test is the combination of the individual SVE and AS tests. Helium as a tracer gas would not be necessary in the combined test.

Table 5.7 Typical Field Data Sheet for AS Pilot Test — Part I

Air Sparge Test Field Data - Part I

Site: ____________________

Location: ____________________

Job: ____________________

Site Personnel: ____________________

Date	Time	Hours	Sparge Wells SW-1								
			Air Line		Helium Line		Combined Line			DTW (feet)	DO(GW) ppm
			Pressure (psi)	Flow (cfm)	Pressure (psi)	Flow (cfm)	Pressure (psi)	Flow (cfm)	Temp. F		

Table 5.7 Typical Field Data Sheet for AS Pilot Test — Part II

Air Sparge Test Field Data - Part II

Site: __________
Location: __________
Job: __________
Site Personnel: __________

Date	Time	Hours	Monitoring Wells MW-1						Monitoring Wells MW-2					
				Helium conc.		VOC	DO	Induced		Helium conc.		VOC	DO	Induced
			DTW (feet)	Vadose %	Sat. %	Vadose ppm	GW ppm	Pressure (in. H2O)	DTW (feet)	Vadose %	Sat. %	Vadose ppm	GW ppm	Pressure (in. H2O)

5.4.3.3 Equipment and Measurement Instruments

Again, basically the same equipment and devices are utilized for the combined test as was utilized for the individual tests. However, if an AS test has already been conducted, helium and DO measurements will not be necessary.

5.4.3.4 Data Collection

Before conducting the combined SVE/AS test, the engineer should evaluate the preliminary data collected from both individual tests to determine the optimum conditions for the combined test. Optimization will be based on the optimum applied vacuum versus flowrate and estimated ROI from the SVE test, and the optimum sparge pressure based on AS ROI generated.

1. Place appropriate wellhead assemblies on extraction, sparge, and observation wells. The fittings should enable measurements of water levels during the test.
2. Initiate SVE test first.
3. Record the following information:

At extraction well:	applied vacuum
	flowrate
	depth to water
At observation wells:	induced vacuum
	flowrate
	depth to water

4. Initiate AS test after steady-state conditions have been achieved with the SVE system.
5. Record the following information along with those in (3):

At sparge well:	applied pressure
	flowrate

6. If positive pressure responses are observed, increase the extraction vacuum rather than decrease the sparge pressure.
7. Collect at least three air samples (both before and after treatment):

   toward end of SVE test
   at the beginning of combined test
   toward end of combined test

No helium or VOC field measurements are made at the observation wells because the venting effect would render those readings meaningless. Table 5.8 provides an example of a field test data sheet.

Table 5.8 Typical Field Data Sheet for AS-SVE Pilot Test

Air Sparge - Soil Vapor Extraction Test Field Data

Site: ____________________

Location: ____________________

Job: ____________________

Site Personnel: ____________________

Date	Time	Hours	Extraction Well			Sparge Well		Observation Wells					
								MW-1			MW-2		
			Applied Vacuum (in. WC)	Flowrate (cfm)	DTW (feet)	Applied Pressure (psi)	Flowrate (cfm)	Vacuum (in.WC)	Flowrate (cfm)	DTW (feet)	Vacuum (in.WC)	Flowrate (cfm)	DTW (feet)

5.5 DESIGN OF SVE/AS SYSTEMS

The design of an SVE or SVE/AS system usually follows the four-step design procedure, as described in Chapter 4.

5.5.1 System Planning

5.5.1.1 Previous Findings and Conclusions

This process involves the review and analyses of relevant existing reports, including problem assessment reports, pilot test results, and the latest quarterly monitoring reports. Pertinent information to compile include:

- Site geology and hydrogeology
- Well logs
- Contaminant concentrations and distribution
- Remediation systems installed to-date, if any
- SVE or SVE/AS pilot test results
- Latest monitoring reports

5.5.1.2 Define Design Objectives

Before proceeding with the detailed design process, the design engineer must determine the answers to several key questions.

- Is vadose zone remediation feasible?
- Is groundwater contamination cleanup feasible?
- What are the required cleanup goals?
- Which remedial option(s) to use?
- What is the estimated time-frame for cleanup?

5.5.1.3 Identify Design Parameters

Certain parameters will need to be established for design purposes. These include:

- SVE ROI
- AS ROI
- Applied vacuum required for achieving desired SVE ROI
- Air flowrate achieved per SVE well at applied vacuum
- Applied pressure required for achieving desired AS ROI
- Air flowrate achieved per AS well

5.5.2 System Design

5.5.2.1 Detailed Final Design

The detailed design of the remediation system will be based on the previous findings and conclusions, design objectives defined, and design parameters established. The design process can be divided into the following stages:

- Preparation of process instrumentation and control diagrams
- Placement of SVE or SVE/AS wells
- SVE blower size determination
- AS blower/air compressor
- Vapor treatment equipment selection
- Utilities hookup identification
- Wellhead design
- Piping manifold design
- Trenching
- Miscellaneous civil, architectural, and structural design

Design drawings and construction specifications are prepared concurrently with the detailed design process. Typical drawings include site plans; site and remediation compound layout; civil, structural, and architectural drawings; wellhead, trenching, and piping manifold details; process and instrumentation diagrams (PID); and electrical drawings. In general, for a typical corner service station site, a total of eight drawings are generated.

5.5.2.2 Computer Models

There are many computer models available to assist design engineers in determining the appropriateness and potential effectiveness of an SVE system to remediate a specific site. A few of the more well-known models are VENTING, AIRFLOW, AIR3D, Airtest, and HyperVentilate.

Some of the models, such as Airtest, AIRFLOW, and AIR3D, are subsurface vapor transport models that predict flow of soil vapor through porous media as a result of the pressure gradient created by the SVE system. These models are useful in evaluating the soil permeability to air flow that is essential to SVE system design. However, these models do not directly assist design engineers in designing an SVE system.

1. determine whether an SVE system is appropriate at a specific site
2. predict the contaminant removal rate under SVE conditions
3. estimate the number of extraction wells needed

Other models, such as VENTING and HyperVentilate, are more complete SVE screening models. Both VENTING and HyperVentilate were developed based on an article written by Paul C. Johnson et al. (1990) of Shell Oil Company. Both of these models can evaluate the soil permeability, estimate the air flowrate, contaminant residual concentrations, contaminant removal rates, and various SVE system design parameters, such as number of extraction wells needed, cleanup time, and system operating conditions.

Since HyperVentilate is the computer model that the authors are most familiar with, a little in-depth discussion will be devoted to HyperVentilate. Generally, the best way to discuss a computer model is to guide the user step-by-step through the entire modeling process. However, it is not the intent of this book to be a software user manual, so only the general features of HyperVentilate will be briefly discussed.

HyperVentilate is a mouse-driven software guidance system for SVE system design. It will guide the user through a structured decision-making process with illustrations and discussions of the various stages of the program. Its capabilities include:

1. Identify and characterize required site-specific data.
2. Determine the applicability of SVE at the site.
3. Evaluate air permeability test results.
4. Determine system design parameters, such as number of SVE wells required and potential operating conditions.

Upon delineation of the extent of contamination in a specific site, the data collected can be used to screen the appropriateness and effectiveness of an SVE system to remediate the site with HyperVentilate.

In evaluating the appropriateness of an SVE system, HyperVentilate utilizes a decision-making process as illustrated in Figure 5.8 to help the user calculate system air flowrates and maximum vapor concentrations. In these calculations, HyperVentilate prompts the user to input certain site-specific information, such as subsurface temperature, extraction well radius, ROI, soil permeability (user can input this value in a range or select a soil type that corresponds to a specific permeability range), type of contaminants (fresh gasoline, weather gasoline, or input a specific contaminant distribution), and permeable zone thickness.

With this information, the maximum removal rates can be calculated by:

$$\text{maximum removal rate} = \text{flowrate} \times \text{maximum vapor concentration} \quad (5.11)$$

Since HyperVentilate is only a screening tool, all of the calculated results are presented in ranges, usually in increments of one order of magnitude. The maximum removal rates are presented in ranges for a wide range of applied vacuums at the extraction well.

The calculated maximum removal rates at a desired applied vacuum can then be compared to the desired removal rates to evaluate the acceptability of

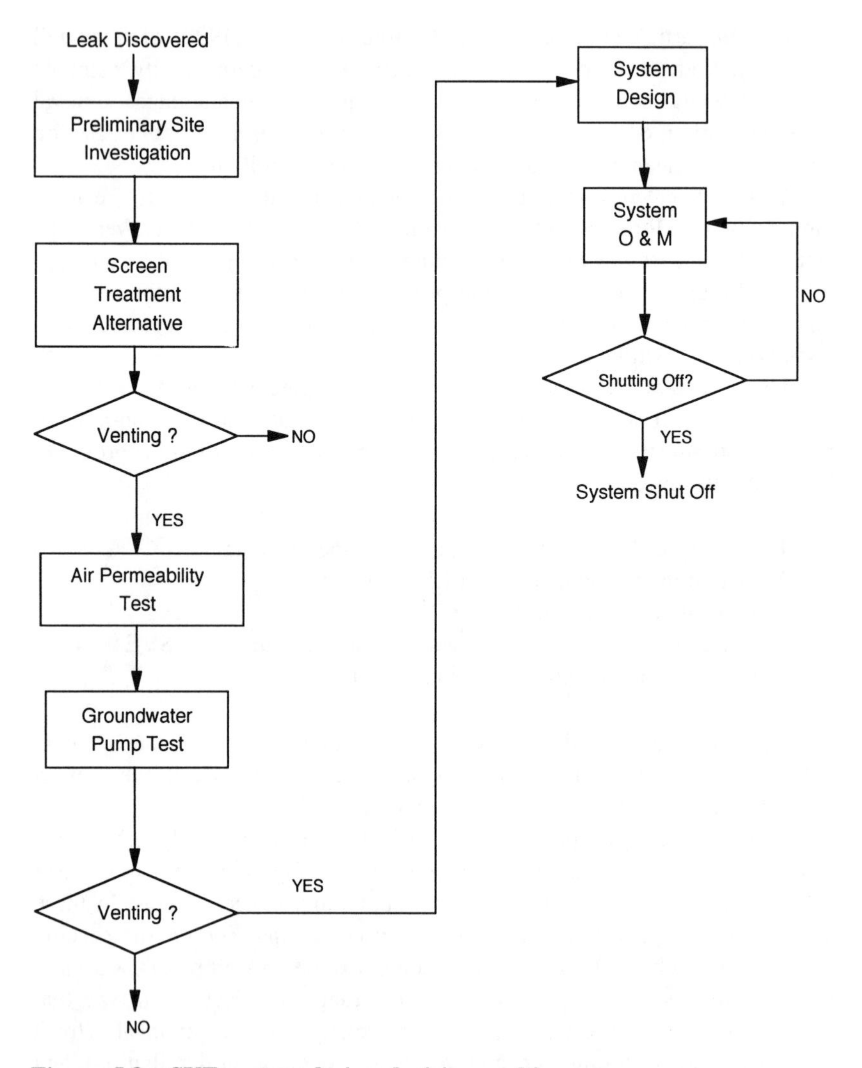

Figure 5.8. SVE system design decision-making process.

the maximum removal rates. If the calculated removal rates are not acceptable, SVE is probably not suitable to remediate the site. If the removal rates are acceptable, the user will then proceed to calculate the first estimate of the minimum number of extraction wells required.

The minimum number of extraction wells required is given by:

$$\text{min. no. of wells} = \frac{(\text{critical volume}) \times (\text{estimated spill quantity})}{(\text{desired cleanup time}) \times (\text{flowrate per well})} \quad (5.12)$$

The critical volume is the amount of air that must be extracted to remove all of the spilled contaminants taking into account that the composition of the contaminant will change with time. HyperVentilate will estimate the value of the critical volume. After a range of the minimum number of wells required is calculated by HyperVentilate, the user will have to decide if this range of wells required is appropriate for the site. If the range of wells required is acceptable, a pilot test should be conducted to collect the field data needed for final system design, such as applied vacuum at the extraction well, extraction well flowrates, extraction well screen intervals, and radial distance of monitoring points from extraction well.

When all of the required data for air permeability calculations are entered into HyperVentilate, it calculates the soil permeability for air flow in Darcy. The calculated soil permeability is then plugged back into the system air flowrates calculation to arrive at revised maximum removal rates and subquently a final estimate of wells required to remediate the site.

The next step is to determine the location and spacing of the extraction wells. The only help HyperVentilate offers in choosing the well location and spacing is a paragraph of text briefly explaining the ideal locations of extraction wells. Basically, the location and spacing of extraction wells should be designed to cover the entire contaminant plume with the minimum number of wells.

5.6 DETAILED DESIGN

5.6.1 Placement and Design of SVE or SVE/AS Wells

The placement and design of SVE and AS wells is probably the most crucial aspect in the design process as far as successful remediation is concerned. SVE wells should be screened such that the flux being drawn to them will pass through the areas of contamination. If there exists distinct layers of contamination, SVE wells with different screen intervals should be installed. There are two major reasons for this. First, a long screen interval may not be effective in inducing airflow near the bottom half of the screen, and second, contaminants trapped in less permeable layers will be remediated at a much slower rate than those found in the more permeable zones.

There are two schools of thought regarding vapor extraction well screen lengths. The predominant view suggests an optimum length of 10 ft, with 15 ft as the maximum. The rationale behind this is that below the first 10 ft of screens, the applied vacuum decreases significantly. The other view takes the practical approach that screen materials are inexpensive, and as such, should be installed across the zone of contamination, regardless of length. In addition, in highly permeable sandy soils, the flux induced across the lower sections of well screens may not be significantly reduced. Based on the author's experience, this observation may be true (HWS Consulting Group, 1995).

Screening vapor extraction wells across heterogeneous zones of contamination always presents a problem. An SVE well screened across a sandy layer and a silt layer will hardly induce any flux through the silt layer. To remediate the silt zone, additional SVE wells screened only in the silt layer will have to be installed.

SVE wells are usually not screened into the water table. The main reason for this is to prevent excessive mounding effects that would reduce the effective available screen interval. Additionally, it will most likely increase the amount of moisture drawn into the system that will have to be removed before entering the blower.

AS wells are screened into the water table and usually have approximately 2 ft of screen. Several factors have to be taken into account in determining the depth of the screen beneath the water table. These include layers or lenses of soils with different permeabilities intermixed below the water table, seasonal fluctuations of the water table, and depth of contamination, if established.

SVE wells should be spaced so as to provide a combined area of influence that is capable of drawing a flux through the area desired for remediation. Where air sparging is utilized to remediate contaminated groundwater, SVE wells should also be located such that off-site migration of air-sparged vapors is prevented.

Sometimes, when the ROI is limited, horizontal SVE and/or AS wells may be considered. However, the cost of drilling horizontal wells is approximately ten times that of vertical wells. When the vadose zone is short, i.e., high water table, horizontal wells can be installed by trenching.

5.6.2 SVE Blower Size Determination

After determining the number of extraction wells needed, the total airflow through the system can be calculated. The next stage in the design process will involve sizing the blower. A review of basic pressure terms and relationships might be appropriate here.

Two scales of pressure measurement exist. Gauge pressure can be either positive or negative, with atmospheric pressure at sea level as its reference. Atmospheric pressure at sea level, which is 14.7 psi, would be zero on the gauge scale. For example, when the reading on the pressure side of a system registers 25 psi, it is showing the excess pressure above atmospheric pressure. Absolute pressure, on the other hand, includes atmospheric pressure in its value. As such, it indicates the pressure above a perfect vacuum (where there are no air molecules to assert any weight at all). In the above example, the absolute pressure would be 25 psi + 14.7 psi, or 39.7 psi (assuming at sea level). To differentiate between gauge and absolute pressure, the units psig and psia are used respectively. Figure 5.9 provides a diagram that illustrates the relationship between gauge and absolute pressure. Absolute pressure must be used in virtually all calculations involving pressure ratios.

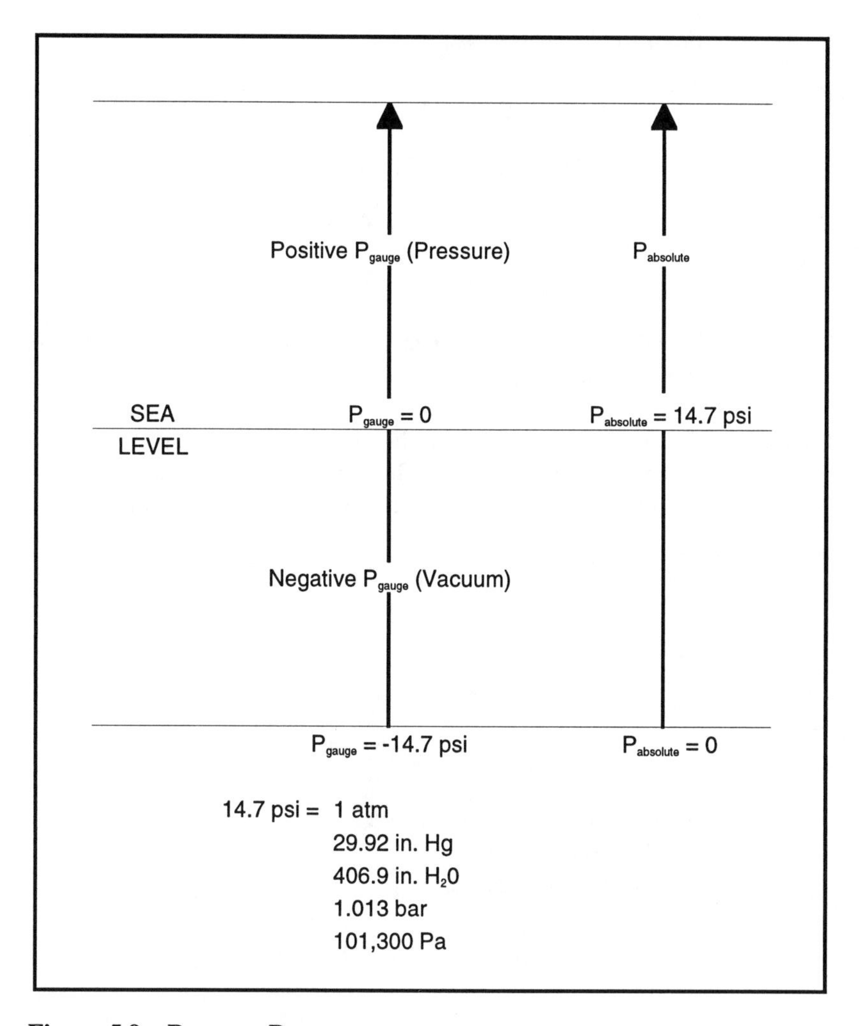

Figure 5.9. P_{gauge} vs. $P_{absolute}$.

In selecting blowers, the term "standard air" will sometimes be encountered. Standard air is defined as air at standard conditions. The following are some examples as to how changes in barometric pressure affects blower design.

Example: At standard air conditions, a blower can handle 500 CFM at 100 in. WC vacuum. If operating at 100 in. WC applied vacuum, what rate of airflow will the blower induce at an SVE site in the mile-high city, Denver?

The altitude in Denver is 5280 ft above sea level. This equates to an atmospheric pressure of 24.63 in. Hg.

From Boyle's Gas Law,

$$P_1V_1 = P_2V_2 \tag{5.13}$$

$$29.92 \times 500 = 24.63 \times V_2$$

$$V_2 = 29.92/24.63 \times 500$$

$$= 607.5 \text{ CFM}$$

Because the air is lighter or less dense at higher altitudes, the same vacuum will be able to draw higher flowrates.

Example: At an SVE site in Denver, the pilot test indicated that a blower operating at 100 in. WC applied vacuum and drawing 500 CFM is required. What standard air blower should be specified?

From Boyle's Gas Law (equation 5.13),

$$P_1V_1 = P_2V_2 \tag{5.13a}$$

$$24.63 \times 500 = 29.92 \times V_2$$

$$V_2 = 412 \text{ CFM}$$

A blower delivering 100 in. WC negative pressure (vacuum) at 24.63 in. Hg is needed.
Therefore, at standard air,

$$\text{blower pressure} = 100 \times 29.92/24.63$$

$$= 121.5 \text{ in. WC}$$

Therefore the blower should operate at 121 in. WC and 412 CFM at standard air.

These examples illustrate that:

1. Operating where atmospheric pressure is lower (higher elevation) will increase the amount of air the blower can induce.
2. Operating where atmospheric pressure is lower will reduce the vacuum the blower can produce. In the previous example, a blower with a nominal rating of 121 in. WC at standard air can only produce an adjusted vacuum rating of 100 in. WC at 5280 ft altitude.

5.6.3 AS Blower/Air Compressor Selection

The AS blower/air compressor needs to be sized such that it can provide the injection pressure necessary to depress the water column above the well screens in all of the AS wells. Since the injection pressure can be significantly higher during seasonal high water tables, this factor should be taken into account when sizing the blower or compressor. Likewise, pressure drops should be calculated where there are significant lengths of piping involved.

Generally, there are three types of AS blower/air compressors that are commonly used in AS systems. They are

1. Rotary lobe blowers can pressurize air up to 15 psi with high air flowrate.
2. Reciprocating air compressors can produce high pressure but low air flowrate.
3. Regenerative blowers are relatively low-pressure blowers.

5.6.4 Piping Design

5.6.4.1 SVE/AS Piping

Piping for SVE and AS systems is designed to be able to withstand the highest anticipated air pressure. This is especially applicable for AS pipes. Generally PVC pipes are used, with schedule 40 for underground installations and schedule 80 ultraviolet (UV)-resistant pipes for aboveground installations. Piping materials should be compatible with the type of contaminant.

Based on length of piping, head loss calculations can be made. The smaller the diameter of the pipe, the greater the head loss. Head loss graphs, tables, and nomographs for pipes of various sizes can be obtained from most blower, compressor, and flow measurement equipment manufacturers and from books on pipings.

Both SVE and AS piping manifolds should be designed to allow for individual well monitoring and control. Figures 5.10 and 5.11 show examples of SVE and AS piping manifolds, respectively.

5.6.4.2 Natural Gas Piping

Where the vapor abatement system requires supplementary fuel, natural gas, if available, is usually provided. Natural gas supplied to most residences and small businesses comes at a low pressure, approximately 0.3 psi or 8 in. WC. Typically, a thermal oxidizer requires a medium pressure of approximately 2 psi. The design engineer's first duty is to inquire with the local gas company if the required pressure is available at the specific location. Piping from the service meter to the unit will have to be included in the system. The size and material for the piping is specified in the Uniform Plumbing Code (UPC; 1994), Chapter 12: Fuel Gas Piping.

Manufacturers of thermal oxidizers provide supplementary fuel criteria in their equipment specifications. Based on these specifications, and the length of piping from the meter to the unit, the diameter of the piping can be determined from tables provided in Chapter 12 of the UPC.

5.6.4.3 Trench Backfill Material

The engineer should also specify in the specifications the backfill material to be used and the method of compaction. Pipe trenches are typically backfilled with noncohesive materials such as sand, aggregate, and pea gravel. When backfilling with well-graded material such as Class II aggregate or well-graded sand, a standard procter test (ASTM D 698) or a modified procter test (ASTM D 1557) should be performed on the material.

The material is laid in lifts between 8 and 12 in. and compacted to at least 95% of the maximum dry density established in the procter test. The percentage of compaction depends on the type of surface overlying the trench and its use, and also whether the standard or modified procter test was specified. In general, the percentage compaction required is lower for the modified test because higher compaction percentages are easier to achieve with the standard test.

Field compaction tests can be accomplished using in-place nuclear tests (ASTM D 2992). If free-draining soil, i.e., non-well-graded material, such as uniform sand or pea gravel, is used density tests (ASTM D 4253 and 4254) are generally used to establish field compaction requirements. In situ nuclear tests are used to measure field compaction achieved. With relative density tests, the percent compaction achieved is typically 60 to 75% of the relative density. Figure 5.12 is a cross-section of a typical SVE/AS trench.

5.6.5 Vapor Treatment Equipment Selection

In most states, treatment of the off-gas is required by the regulations. The air permit usually specifies a destruction removal efficiency (DRE) and limits for emission rates (daily, monthly, or annually).

The four most common methods for treating vapor gas generated by SVE systems are

1. Carbon adsorption
2. Thermal oxidation
3. Catalytic oxidation
4. Internal combustion engines (ICE)

Other vapor abatement technologies that have been used include biofilters and ceramic bead beds. Figure 5.13 illustrates the different ranges of contaminant concentrations that are applicable for each type of treatment option (U.S. EPA, 1990a).

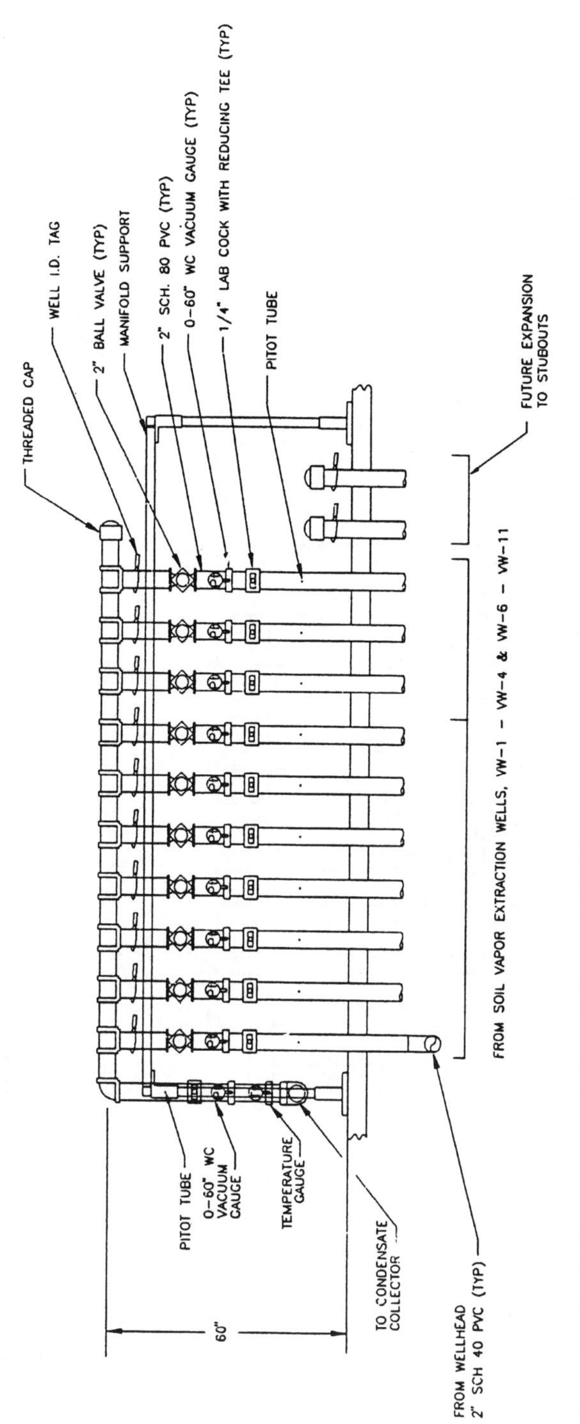

Figure 5.10. Soil vapor extraction piping manifold system.

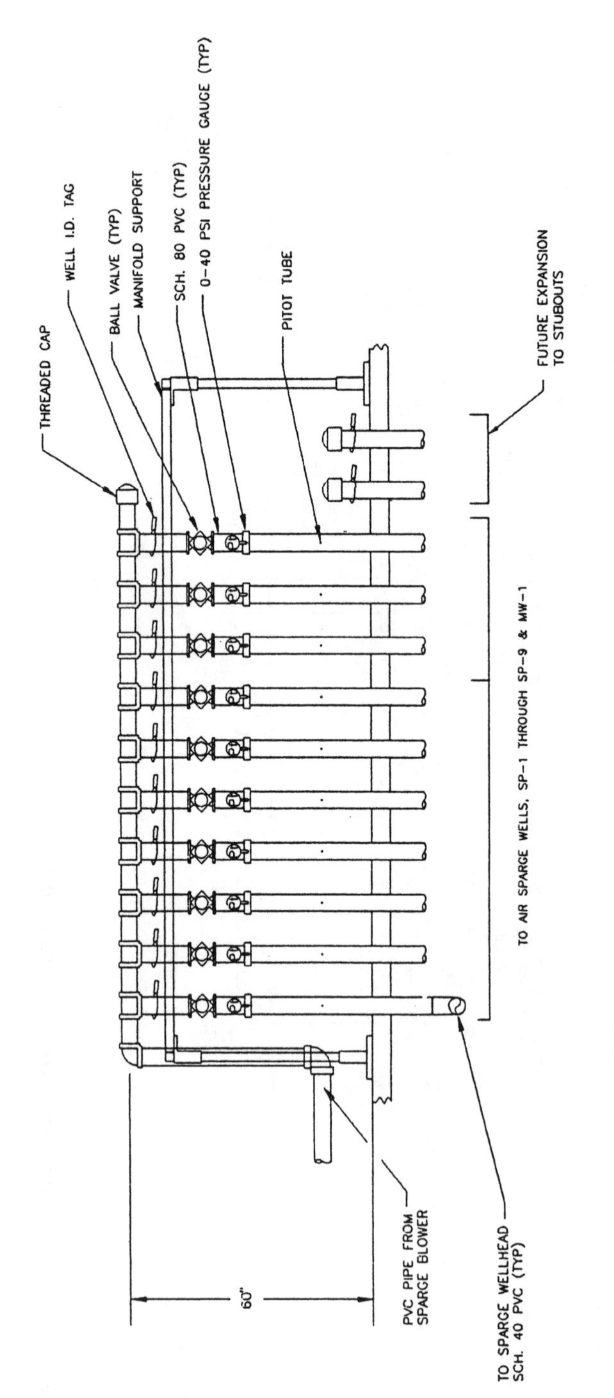

Figure 5.11. Air sparge piping manifold system.

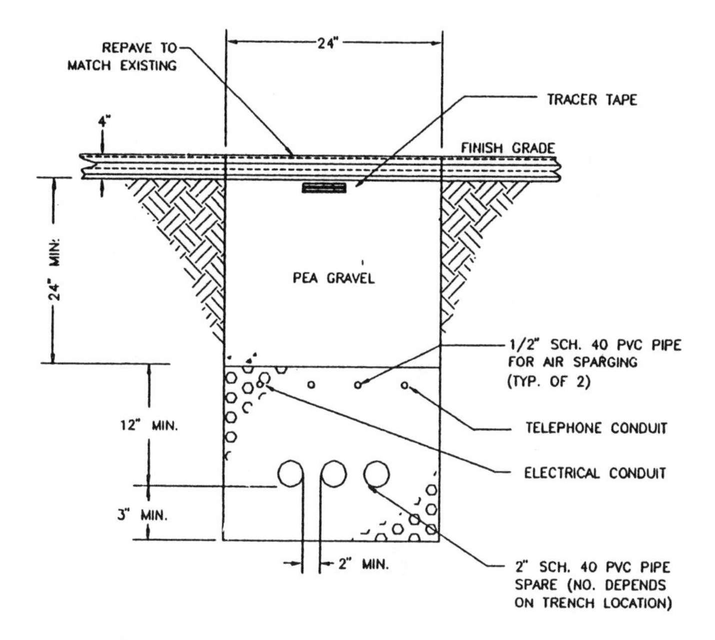

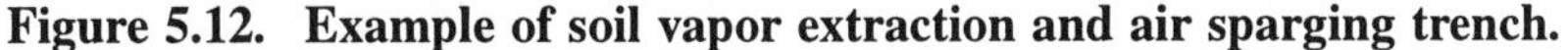

Figure 5.12. Example of soil vapor extraction and air sparging trench.

5.6.5.1 Carbon Adsorption

Carbon adsorption is a conventional technology for the removal of VOC from a gas stream. In the treatment of vapor extraction off-gas, it has the capability to handle variable influent conditions, i.e., hydrocarbon concentrations and flowrates. However, when the concentrations are high, the cost of replacing the carbon becomes cost-prohibitive. As such, carbon adsorption is only used when the influent concentrations are low or have stabilized to a cost-effective level.

Activated carbon is used as an adsorbent material because its porous structure provides a large surface area. As a rule of thumb, carbon can adsorb approximately 10 to 40% of its own weight, thus, for every 10 to 40 pounds of hydrocarbons removed, approximately 100 pounds of carbon is required. Granular activated carbon can either be regenerated or disposed of and replaced. The moisture content of the process stream is an important consideration in carbon adsorption treatment. The adsorbed moisture essentially restricts access to active sites on the carbon's surface and therefore reduces its capability.

5.6.5.2 Thermal Oxidation

Thermal oxidation refers to the burning or incineration of hydrocarbon-laden vapors; in the process, VOC react at high temperatures with oxygen to

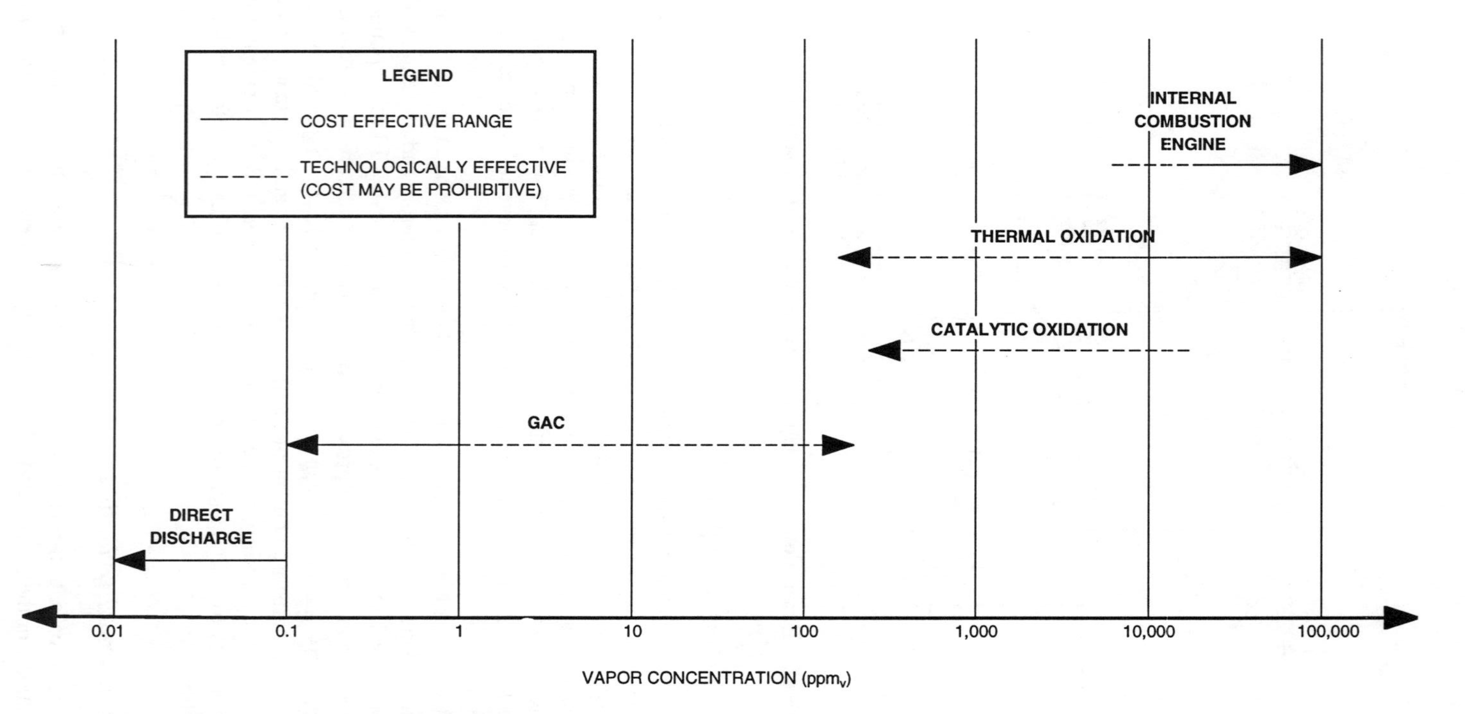

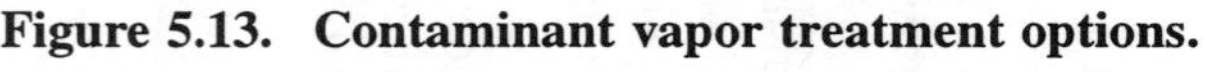

Figure 5.13. Contaminant vapor treatment options.

Table 5.9 Flammability Limits for Selected Organics in Air

Volatile Organic Compounds	LEL (% by Volume)	UEL (% by Volume)
Acetaldehyde	4.0	6.0
Acetone	2.6	12.8
Benzene	1.3	7.1
Carbon Disulfide	1.3	50.0
Chlorobenzene	1.3	7.1
Cyclohexane	1.3	8.0
Hexane	1.1	7.5
Methane	5.4	15.0
Methyl Cyclohexane	1.2	6.7
Toluene	1.2	7.1
Trichloroethylene	12.5	90.0
Xylene, m- and p-	1.1	7.0
Xylene, o-	1.0	6.0

form carbon dioxide and water while releasing heat. The thermal oxidation unit operates at temperatures typically in the 1200°F to 1600°F range. Most units are sold as complete systems, and come with a vapor extraction blower, a moisture knockout drum, the combustion chamber, an air dilution system, the exhaust stack, and automatic controls.

Thermal oxidation units can handle high concentrations and are generally more appropriate for startups and the initial phases of remediation when influent concentrations are typically much higher. However, regulatory agencies may restrict the maximum allowable VOC concentrations, usually as a percent of LEL because of safety reasons. The flammability limits for selected organics in air at standard pressure and temperature are provided in Table 5.9.

Because of the high hydrocarbon concentrations and the natural biological activities in the soil, it is common to encounter low oxygen levels and high carbon dioxide levels in the extracted vapors. As such, the process gas is diluted with ambient air to ensure sufficient oxygen for combustion. Ambient dilution air is drawn into the system at the suction side of the blower. Flow control valves operate together to ensure that the ratio of dilution air to process air can be varied so as to keep the total flow constant. Because of the increase in dilution air (contaminant concentrations are now decreased), supplemental fuel has to be added to the system.

Natural gas (methane) or propane are the typical sources of auxiliary fuel. Auxiliary fuel is also needed after the initial high VOC concentration levels have settled down to lower, more steady-state conditions. Anytime that the temperature in the oxidizer falls below a preset temperature that is required to completely destroy the VOC, the system will attempt to add supplemental fuel. If the temperature rises above the upper preset limit, fuel will be reduced and dilution air increased.

The performance of incineration is dependent on three parameters: temperature, residence time, and turbulence (the "three Ts"). Danielson (U.S. EPA, 1990a) suggested that, for efficient destruction, incinerators should be designed to operate at temperatures of 800 K to 1,100 K (1260°F to 1560°F), with residence times between 0.3 to 0.5 seconds, and flow velocities of 6 to 12 m/s. The residence time is important to ensure that the reactions have sufficient time to reach the desired degree of completion. Adequate velocity is desired to promote turbulent mixing (the third "T") in the incinerator. Turbulence provides for adequate mixing of VOC and oxygen during the process.

VOC destruction rates are extremely sensitive to temperature. To ensure complete combustion and, thus, comply with the requirements of the air discharge permit, the combustion temperature has to be maintained between the limits specified by the manufacturer. To stay within the preset temperature range, the manufacturer specifies the heat input rating (in BTU/hr) for the particular model. For a 100 ACFM unit, typical ratings are between 500,000 to 1,000,000 BTU/hr. To ensure that adequate heat energy is provided to the oxidizer, the manufacturer also provides recommended influent VOC concentration ranges, in addition to operating temperature requirements.

If the VOC content in the process gas (air stream from extraction wells), is less than the lower range, it will not be economical to operate the particular system because of the quantity of supplemental fuel needed to maintain the temperature required to completely destroy the VOC. If the VOC content is higher than the upper range, too much heat will be generated in the oxidizer and could potentially damage the unit. In this case, the ratio of dilution air to process gas will be increased, thereby reducing the VOC concentration. During the design process, it is necessary to determine that the natural gas service available is at the manufacturer's specified pressure. If propane fuel is used instead, check with the manufacturer on the minimum tank size required.

In incineration, the stoichiometry for complete combustion is

$$C_xH_y + (x + y/4)O_2 \rightarrow xCO_2 + (y/2)H_2O$$

In the operation of a thermal oxidizer for the incineration of VOC vapors, two oxidation reactions take place:

VOC from extracted vapor (toluene, C_7H_8, as an example):

$$C_7H_8 + 9O_2 \rightarrow 7CO_2 + 4H_2O$$

Methane, CH_4, from auxiliary fuel:

$$CH_4 + 2O_2 \rightarrow CO_2 + 2H_2O$$

When using oxidizers as air abatement devices, numerous calculations are required. These include calculations for the application-to-construct (ATC) permit, permit-to-operate (PTO) application, and quarterly reporting data for the discharge permit. Engineering calculations for the design of the oxidizer itself are typically performed by the manufacturer.

Basic Equations/Calculations for Air Calculations

1. Conversion from mg/L to ppm_v and vice versa. The conversion is based on the Ideal Gas Law,

$$pV = nRT \tag{5.14}$$

where
p = pressure (atm)
V = volume (L)
n = number of moles
R = constant (atm-L/mole-K)
T = temperature (K)

At standard temperature and pressure (STP), the volume of 1 mole of ideal gas is 22.4 L.

Therefore,

$$R = \frac{(1 \text{ atm})(22.4 \text{ L})}{(1 \text{ mole})(273 \text{ K})}$$

$$= 0.0821 \text{ atm-L/mole-K}$$

The value of R is the same for all gases since at STP all gases have the same number of molecules in one mole (6.02×10^{23} molecules).

From equation (5.14),

$$n = PV/RT$$

$$\text{mole} = \frac{\text{atm x L}}{0.0821 \text{ x K}} \tag{5.15}$$

Multiply both sides by mg/L and divide both sides by mole,

$$\text{mg/L} = \frac{\text{atm x mg/mole}}{0.0821 \text{ x K}} \tag{5.16}$$

The dimension mass/mole describes the molar mass of a material and is given in g/mole. Where x is the formula mass of a material, its molar mass is x grams per mole of the substance. It is thus equivalent to its molecular weight. Example, the formula mass for benzene, C_6H_6, is (12 × 6) + (1 × 6) = 78. Thus, its molar mass is 78 g/mole or 78,000 mg/mole. The molar mass for gasoline is 95 g/mole or 95,000 mg/mole.

At STP,

$$\text{mg/L} = \frac{(1\text{ atm})(95{,}000\text{ mg/mole})(\text{ppm}_v)}{(1{,}000{,}000)(0.0821)(273\text{ K})} \tag{5.17}$$

where the figure 1,000,000 is for conversion from ppm_v.

Therefore, from the above equation, for gasoline, when the concentration is 1 ppm_v, the equivalent concentration in mg/L is 0.00424 mg/L.

Similarly, from the above equation, for gasoline, when the concentration is 1 mg/L, the equivalent concentration in ppm_v is 235.9 ppm_v.

2. Extraction and discharge rates (lb/day) and destruction efficiencies. Extraction rates refer to the rate (in lb/day) of hydrocarbon vapor removal from the subsurface. It typically measures TPHg (or TRPh in some states) and benzene. The formulae is given by:

$$\underset{(\text{lb/day})}{\text{Extraction Rate}} = \underset{(\text{CFM})}{\text{Flowrate}} \times \underset{(\text{mg/L})}{\text{Influent Conc.}} \times 0.088 \tag{5.18}$$

OR

$$\underset{(\text{lb/day})}{\text{Extraction Rate}} = \underset{(\text{CFM})}{\text{Flowrate}} \times \underset{(\text{ppm}_v)}{\text{Influent Conc.}} \times 0.00037 \tag{5.19}$$

Likewise,

$$\underset{(\text{lb/day})}{\text{Emission Rate}} = \underset{(\text{CFM})}{\text{Flowrate}} \times \underset{(\text{mg/L})}{\text{Influent Conc.}} \times 0.088 \tag{5.20}$$

OR

$$\underset{(\text{lb/day})}{\text{Emission Rate}} = \underset{(\text{CFM})}{\text{Flowrate}} \times \underset{(\text{ppm}_v)}{\text{Influent Conc.}} \times 0.00037 \tag{5.21}$$

The efficiency of the oxidizer is measured by the hydrocarbon destruction efficiency, given by:

$$\text{Destruction Efficiency} = \frac{\text{Extraction Rate} - \text{Emission Rate}}{\text{Extraction Rate}} \times 100\% \quad (5.22)$$

In the preparation of ATC, maximum extraction rates are typically required by the agencies regulating the discharge permits. Because it can be difficult, depending on site conditions and available data, to reliably estimate (from soil and groundwater concentrations) or determine (from SVE tests) the maximum extracted hydrocarbon vapor concentrations, the maximum extraction rate can be calculated based on the maximum system capacity of the oxidizer. The manufacturer typically specifies the maximum influent concentration that the unit is designed to handle without the temperature rising too high. Thus the maximum extraction rate for the particular oxidizer is

$$\underset{(\text{lb/day})}{\text{Max Extraction Rate}} = \underset{(\text{CFM})}{\text{Flowrate}} \times \underset{(\text{ppm}_v)}{\text{Max Influent Conc.}} \times 0.00037 \quad (5.23)$$

3. Conversion of SCFM to ACFM. The volume of air removed (as in SVE systems) or delivered (as in AS systems) by a blower is given in cubic feet per minute (CFM). This may be either CFM at actual temperature and pressure, actual cubic feet per minute (ACFM), or at standard temperature and pressure, standard cubic feet per minute (SCFM). ACFM refers to the actual quantity of air moved, taking into consideration the local effects of temperature and pressure. SCFM refers to the flowrate measured at atmospheric pressure and a standard temperature of 70°F (some manufacturers use 68°F). (For utmost accuracy, SCFM also requires correction to a standard humidity of 36 percent.)

The relationships of volume, pressure, and temperature of a quantity of air are interrelated and expressed by the gas laws.

$$\text{Boyle's law:} \quad P_1V_1 = P_2V_2 \quad (5.24)$$

$$\text{Charles law:} \quad \frac{P_1V_1}{T_1} = \frac{P_2V_2}{T_2} \quad (5.25)$$

$$\text{Combined gas law:} \quad \frac{P_1V_1}{T_1} = \frac{P_2V_2}{T_2} \quad (5.26)$$

From the combined gas law, ACFM and SCFM can be interchangeably converted.

$$\frac{P_a V_a}{T_a} = \frac{P_s V_s}{T_s} \tag{5.27}$$

where P_a = actual pressure (psia)
V_a = actual flowrate (ACFM)
T_a = actual temperature (temperature, °F + 460)
P_s = standard pressure (14.7 psia)
V_s = standard flowrate (SCFM)
T_s = standard temperature (70°F + 460)

Flowrate Calculations

The flowrate into the combustion chamber (influent flow) consists of either process gas (hydrocarbon-laden vapors from the extraction wells) or process gas mixed with ambient dilution air. Control valves on the process gas piping and dilution air piping are linked together to ensure that the influent flowrate remains the same, i.e., at the rated flowrate for the particular oxidizer. Typical nominal flowrates are 100, 200, and 500 CFM. Because of the increase in temperature in the oxidizer chamber, the flowrate increases. Likewise, the flowrate in the exhaust (stack) piping will be greater than the influent flowrate because of the higher temperatures. The exhaust temperature is generally lower than the oxidizer chamber temperature by approximately 500°F to 800°F.

Besides temperature change, other factors that affect flowrates through the thermal oxidizer include supplemental fuel flowrates and mole expansions. Supplemental flowrates are typically very low compared to the process stream and as such are negligible in determining chamber and stack flowrates. For the components of gasoline, the moles of stoichiometric combustion products are usually slightly greater than the input moles. These increases are negligible when calculating oxidizer and chamber flowrates.

(1) Flowrate in oxidizer chamber

$$= \text{Influent Flowrate (CFM)} \times \frac{\text{Chamber Temp (°F)} + 460\text{°F}}{\text{Influent Temp (°F)} + 460\text{°F}} \tag{5.28}$$

(2) Stack Flowrate

$$= \text{Influent Flowrate (CFM)} \times \frac{\text{Stack Temp (°F)} + 460\text{°F}}{\text{Influent Temp (°F)} + 460\text{°F}} \tag{5.29}$$

Typical temperature ranges encountered are

Influent:	50 to 70°F
Combustion:	1200 to 1600°F
Stack:	800 to 1000°F

Velocity Calculations

1. Chamber Velocity (fpm)
   = Chamber Flowrate (CFM)/Chamber Cross-sectional Area (ft^2) (5.30)

2. Stack Velocity (fpm)
   = Stack Flowrate (CFM)/Stack Cross-sectional Area (ft^2) (5.31)

Residence Time Calculations

Chamber Residence Time (min)
= Chamber Length (ft)/Chamber Velocity (fpm) (5.32)

Energy Consumption Calculations

Energy consumption calculations typically involve determining the amount of auxiliary fuel required. The manufacturer of the oxidizer will provide the BTU requirements of the particular unit. From the flowrate and concentration of the process flow and the BTU content of weathered gasoline, calculations can be performed to determine if supplemental fuel is necessary. On-site, control valves and sensors on the unit will automatically determine when auxiliary fuel is required and by how much. Calculations are performed to provide an idea of the operations and maintenance cost of the system insofar as supplemental fuel is concerned.

5.6.5.3 Catalytic Oxidation

Catalytic oxidation is essentially a flameless combustion process that deploys a catalyst to accelerate the oxidation process. Catalysts are precious metals that are coated on an inert structure. The main operating differences between a catalytic oxidizer and a thermal oxidizer are the influent hydrocarbon concentrations and the operating temperature. In both cases, the catalytic oxidizer operates at lower numbers. Hydrocarbon concentrations are typically limited to below 25% of the LEL (approximately 3000 ppm_v). The operating temperature is usually kept in the 700 to 1000°F range. Typically, 650°F is the minimum inlet operating temperature to the catalytic cell and 1200°F is the maximum outlet temperature exiting the catalytic cell.

As with the thermal oxidizer, both dilution air and supplemental fuel are provided. At the maximum temperature, the maximum soil vapor VOC concentration that can be treated without dilution is usually set equivalent to approximately 25% LEL or 3000 ppm_v, measured as hexane. VOC concentrations entering the chamber is continuously monitored with an LEL detector. If the concentration exceeds the percent LEL set-point, dilution air is automatically added to maintain the LEL entering the combustion chamber at the proper concentration. The catalytic oxidizer also has a preheater before the catalyst bed in its combustion chamber. The temperature of the incoming process stream is raised to 550 to 650°F. The preheater may run on electricity or may be gas-fired. In terms of operation, the advantage of having electric preheat is a lesser sensitivity to low oxygen concentrations in the incoming stream.

Variables affecting the performance of a catalytic oxidizer include gas characteristics, catalyst properties, particulates in the gas stream, operating temperature in and out of the catalyst bed, and space velocity, which is the total volumetric flow entering the catalyst bed divided by the bed's volume. Extracted vapor entering the treatment unit should be analyzed for lead content. Lead "blinds" or coats the catalyst and destroys it. If lead is present in the soil (due to leaded gasoline), an alternate treatment system may need to be considered.

5.6.5.4 Internal Combustion Engines

Internal combustion engines are simply automotive engines modified to run on petroleum vapors instead of liquid fuel. Provisions are made for both supplemental fuel addition and ambient air intake to sustain combustion. The ICE unit uses the engine itself as a source of vacuum for drawing vapor into the system. The supplemental fuel is also used for startup of the system and initial operations.

5.6.5.5 Biofilters

Biofilters employ the use of hydrocarbon-degrading microorganisms to break down VOC-laden vapors. The biofilter is usually a structure designed to house the microbes in a controlled environment where suitable amounts of oxygen, nutrients, and moisture are provided.

5.7 DESIGN EXAMPLES

5.7.1 Example 1

The following example assumes all site characterization work, including detailed hydrogeological investigations and pilot tests have already been performed. The design engineer's task is to analyze the available data and prepare a conceptual design.

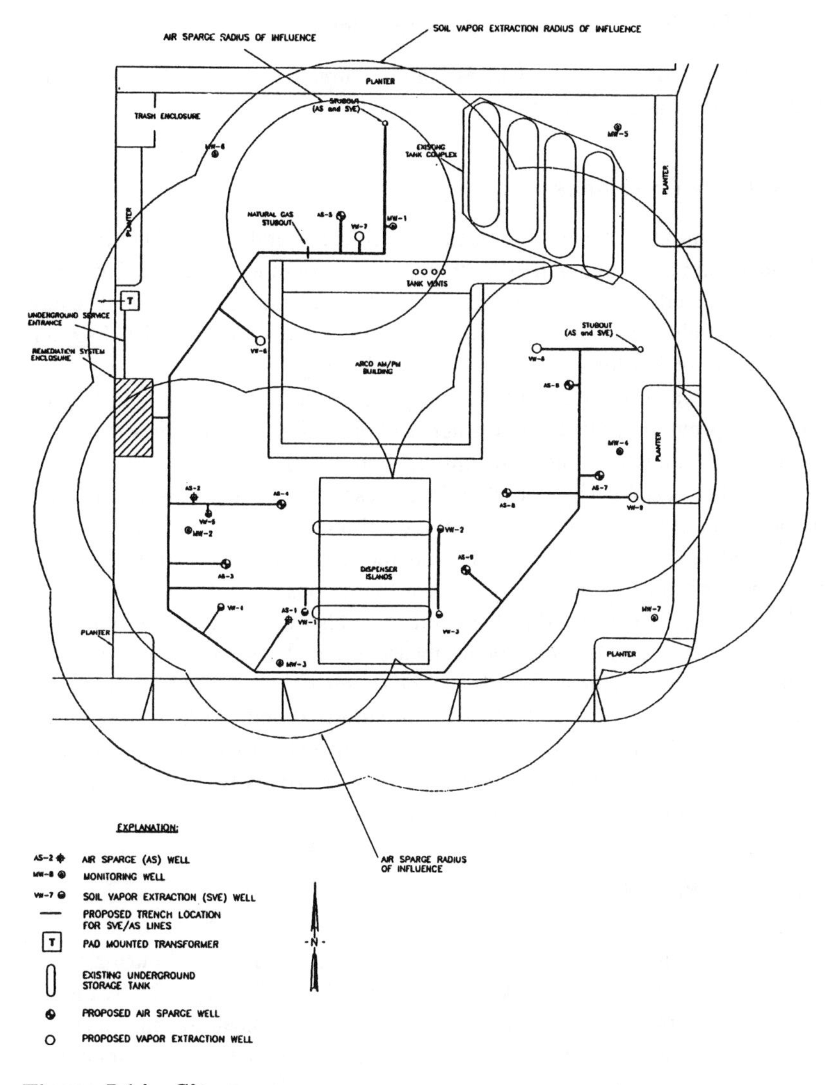

Figure 5.14. Site map.

5.7.1.1 Scenario

The site is a typical corner service station, as shown in Figure 5.14. Underground storage tanks (UST) were removed from the southwest corner of the site, along with product pipings to the dispenser islands. Petroleum hydrocarbon-contaminated soils were discovered during tank removals. Contaminated soils were removed around the tank complex up to a depth of 12 to 15 ft below surface grade (bsg).

5.7.1.2 Previous Findings and Conclusions

A review of studies and investigations already completed provided the following data and information.

5.7.1.3 Site Conditions

Site geology consists of silty sand with lenses and interbedded layers of silt, sand, and clay irregularly distributed throughout the silty sand. The lenses and interbedded layers vary in thickness and lateral extent and occur at several depth intervals: generally less than 2 ft thick between 10 and 17 ft bsg, and approximately 3 ft thick between 25 and 33 ft bsg.

Groundwater has generally fluctuated between 37 and 41 ft bsg in on-site wells the previous year. Groundwater occurs in silty sand to fine- to coarse-grained sand units and under unconfined conditions.

Residual gasoline hydrocarbons are present in soil beneath the site. The highest concentrations of gasoline hydrocarbons were detected in the south-west quadrant of the property. This was the location of the former UST and associated product pipings. Highest concentrations were found between 15 and 30 ft bsg. TPHg and BTEX in soil were also detected on the remainder of the property except around the northeast corner. These concentrations were relatively lower than those recorded in the southeast section and were detected between 27 and 40 ft bsg. Except for monitoring wells MW-5, MW-6, and MW-7, located near the northeast corner, northwest corner, and the southeast corner of the site, respectively, dissolved hydrocarbons were detected in the remainder of the on-site monitoring wells.

5.7.1.4 SVE/AS Pilot Test Results

1. SVE Pilot Test. Applied vacuums of 10 and 20 in. WC during the pilot test resulted in extracted air flowrates of 25 and 42 SCFM, respectively. This is indicative of good air permeability in the soil matrix as significant air flows were obtained at relatively low applied vacuums. Based on induced vacuums recorded in the surrounding observation wells during the test, a vacuum versus distance graph was generated, yielding an estimated ROI of 50 ft.

   Concentrations of total petroleum hydrocarbons as gasoline (TPHg) in extracted vapors during the SVE tests settled at around 22,000 parts per million by volume (ppm_v). Initial hydrocarbon removal rates for the SVE system were arbitrarily estimated to be 325 pounds per day per well, based on a flowrate of 40 SCFM. These initial removal rates typically decrease rapidly with time.

   Soil vapors extracted contained 1.3% oxygen and 15.8% carbon dioxide, and 73 ppm_v of methane. The low levels of oxygen and high levels of carbon dioxide and methane suggest that petroleum hydrocarbons may be biodegrading in the subsurface. The introduction of oxygen through AS may further enhance in situ biodegradation of hydrocarbons in the soil and groundwater.

Significant groundwater mounding, approximately 5 ft, was observed in the extraction well during the test (the well was screened almost into the static groundwater table).

2. Air Sparge Test. After conclusion of the SVE test, an AS test was performed at a nearby sparge well, with screen intervals at 50 to 52 ft bsg. AS test results indicated that air pressures in excess of 150% over the calculated hydrostatic head of 5.2 psig (static water table was at 38 ft bsg) were necessary to induce air flowrates of 3 to 5 SCFM. These air entry pressures are relatively low, indicating that the saturated soils are comprised of relatively permeable soils at the test depths. This observation correlates to the geology observed during drilling activities.

   The ROI determined from AS testing was interpreted based on changes in DO and water levels, and positive pressures measured in the observation wells. In addition, helium was injected as a conservative tracer. These techniques collectively indicated an AS ROI of approximately 30 ft.

3. Combined SVE/AS Test. A combined SVE/AS test was performed after completion of the individual SVE and AS tests. At applied vacuums of 15 and 20 in. WC, a steady induced air flowrate of 30 SCFM was observed. The sparge pressure was slowly built up to and maintained at 9 psig, yielding flowrates of 2 to 3 SCFM.

   The estimated ROI for the extraction well at an applied vacuum of 20 in. WC was approximately 45 ft, irrespective of the AS flow rate. AS thus reduced the SVE ROI for the well from 50 to 45 ft. The AS ROI was estimated to remain at 30 ft.

   Contrary to the individual air sparge test conducted earlier, vacuum (negative) responses were recorded at the observation wells during the combined test as opposed to positive pressure readings. Additionally, the vacuum readings were correspondingly lower than those recorded during the individual SVE test. These observations are in line with what one would expect and provide evidence that the mechanisms of SVE and AS are acting together in creating an effective SVE/AS system. In addition, groundwater mounding effects observed around the SVE well during the combined test was less than during the SVE test alone. The observed rise in water level was 3 ft as opposed to 5 ft. This is an important observation because excessive water level increases would decrease the unsaturated zone area available for venting.

   Concentrations of TPHg in extracted vapor during the combined SVE/AS test averaged 30,000 ppm_v and were higher than the concentrations observed during the SVE test. Initial hydrocarbon removal rates based on results of the combined test were approximately 333 pounds per day per well. Despite the higher concentrations, this is almost equal to the removal rate for SVE alone. This is simply due to the reduced flowrate.

5.7.1.5 Design Objectives

1. Achieve soil cleanup via SVE treatment.
2. Achieve groundwater cleanup via AS-enhanced biodegradation and AS-induced volatilization.

5.7.1.6 Design Parameters

1. For SVE wells, an ROI of 45 ft at an applied vacuum of 20 in. H_2O with 30 SCFM induced flowrate per well will be used as a design guideline.
2. At greater than 20 in. H_2O applied vacuum, considerable upwelling may be expected, depending on both depth to water table and screened intervals. It is not known how far the mounding effect extends beyond the extraction well.
3. For AS wells, an ROI of 30 ft at an applied pressure of 9 psi with 3 SCFM of injected airflow per well will be used as a design guideline.
4. The off-gas treatment unit will be a thermal oxidizer.
5. Other considerations: Horizontal wells were initially considered; however, the concept was abandoned due to prohibitive drilling costs.

5.7.1.7 Conceptual Design

The proposed SVE/AS system will utilize nine vapor extraction wells and nine air sparge wells, two blowers, and an abatement device to process the extracted hydrocarbon-laden soil vapors. Figure 5.14 depicts the remediation system layout. Of the nine vapor extraction wells, five are screened between 18 and 33 ft bsg (wells in the southwest quadrant), and the remaining four are screened between 25 and 35 ft bsg. The capture zones for the wells are depicted in Figure 5.14. The sparge wells will be screened between 50 and 52 ft bsg. The combined ROI for the sparge wells are shown in Figure 5.14.

The combined flowrate for the system is estimated to be approximately 250 SCFM. A 250 CFM thermal oxidizer will be used as the abatement device for treating petroleum hydrocarbon-laden soil vapors extracted from the vapor extraction wells. Initial daily destruction rates of hydrocarbons are estimated to be approximately 3,700 lb, based on extraction rates of 250 SCFM and TPHg concentrations of 30,000 ppm_v.

Considerable thought was given to the location of a sparge well (AS-5) in the vicinity of MW-1 (north of the building). This is simply due to its close proximity to both the building and the existing UST. The risk of uncontrolled migration of sparged vapor gas was weighed against the potential benefits of operating a sparge well here. It is believed that by locating an SVE well close to it, potential problems could be prevented.

A process-and-instrumentation diagram (PID) of the system is illustrated in Figure 5.15.

5.7.2 Example 2

As in Example 1, Example 2 assumes all site characterization work, including detailed hydrogeological investigations and pilot tests, have been performed. The design engineer's task is to review and evaluate the available data and prepare a conceptual design.

5.7.2.1 Scenario

The site is a grain elevator constructed in the mid-1950s, as shown in Figure 5.16. During a property transaction, a Phase I followed by a Phase II Environmental Site Assessment discovered the presence of carbon tetrachloride in the soil and groundwater at the site. Carbon tetrachloride was utilized as a grain fumigant on-site during the early years of operation.

5.7.2.2 Site Characterization

The site is underlain by 240 ft of unconsolidated clays, silts, sands, and gravels. Clays and silts are the predominant soil types to a depth of approximately 40 to 50 ft. These are underlain by sand with some gravel, which becomes coarser with depth. The water table occurs at a depth of 110 to 115 ft below the site though, due to changes in topography, it occurs at a depth of 90 to 100 ft east of the site.

The sand frequently contains gravel below the water table, including some thick gravel zones. Two thin clay units occur within the sand and gravel. The first of these clay units occurs at a depth of 130 ft and is referred to as the upper clay layer. The deeper clay layer is found at a depth of 150 ft and is referred to as the lower clay layer. These units vary in thickness from one to seven ft thick in this area, but do not exist east of the elevator.

The sand and gravel are underlain by shale bedrock. The sand and gravel comprise the principal aquifer in the area, and very high capacity domestic, irrigation, and public supply wells have been developed in it. Groundwater flow in the area is generally to the east-southeast.

Carbon tetrachloride is present in the vadose zone in the immediate vicinity of the elevator buildings, and extends to the water table. The maximum concentration measured in the soil is 884.7 μg/kg. Carbon tetrachloride-contaminated

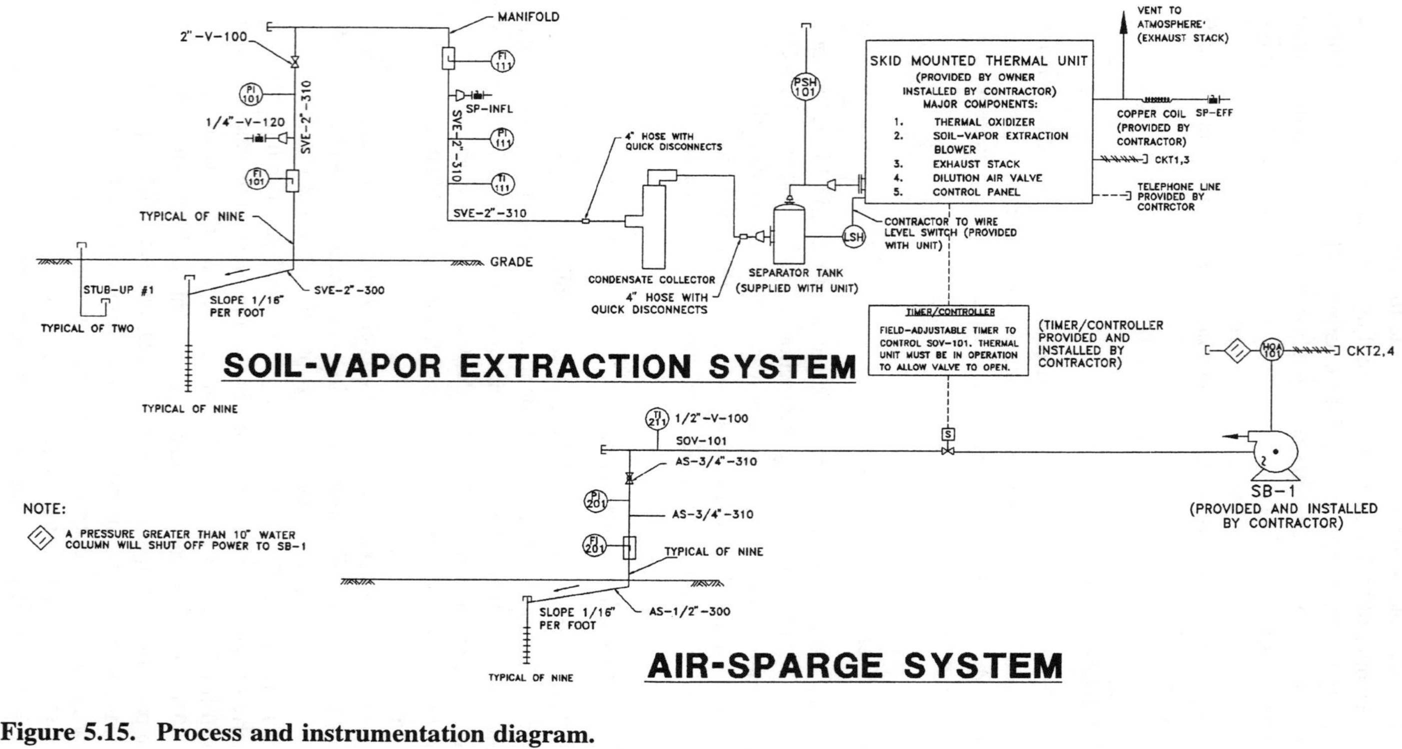

Figure 5.15. Process and instrumentation diagram.

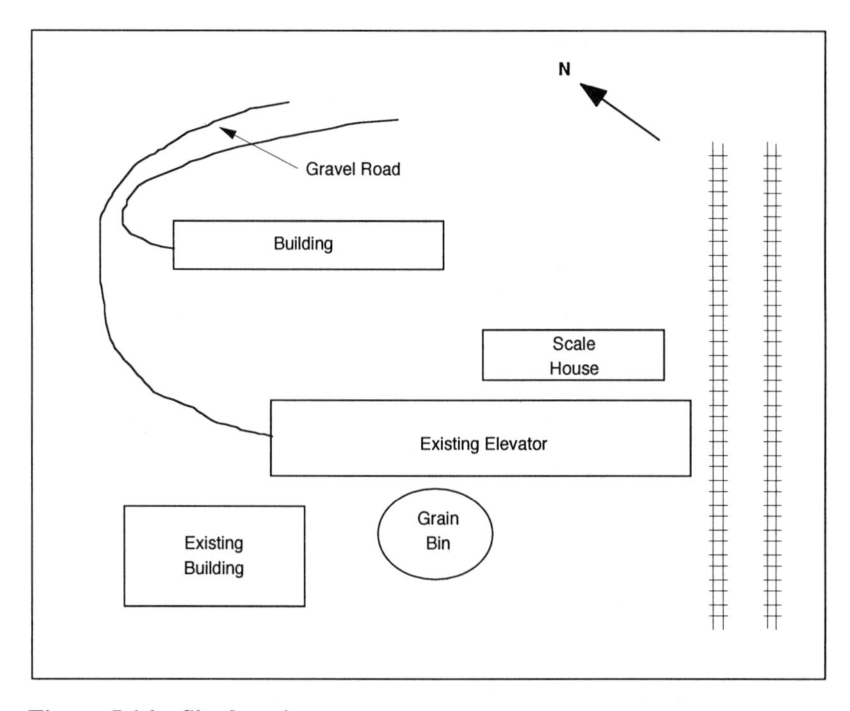

Figure 5.16. Site location map.

soils encompass an area of approximately 500,000 ft^2, and they extend down to the water table. The total soil volume affected is estimated to be 55 million ft^3. Figure 5.17 depicts the extent of contamination in the vadose zone.

Carbon tetrachloride occurs in the groundwater beneath the site at concentrations as high as 29,943 μg/L. The highest concentrations occur in the upper aquifer (above the upper clay layer). Carbon tetrachloride is also present in the zone between the two clay units at the elevator site, though at lower concentrations. No carbon tetrachloride was found in the lower portion of the aquifer or on top of the bedrock in the immediate vicinity of the elevator.

Due to the extent of the contamination and the high costs of remediation, the remedial plan is to concentrate first on source control and removal, then remediation. The remediation options selected are SVE for soils remediation and pump and treat for groundwater remediation. The objective of the SVE system is to control and remove the carbon tetrachloride trapped in the vadose zone that serves as a continuous source of further groundwater contamination. The pump and treat of groundwater is to control further migration of the contaminant plume. Only the SVE portion of the remedial design will be discussed in this example.

5.7.2.3 SVE Pilot Test Results

1. Pilot Test in the Shallow Zone. The shallow SVE well was screened from 10 to 40 ft bgs in the shallow silt and clay zone. The vacuum

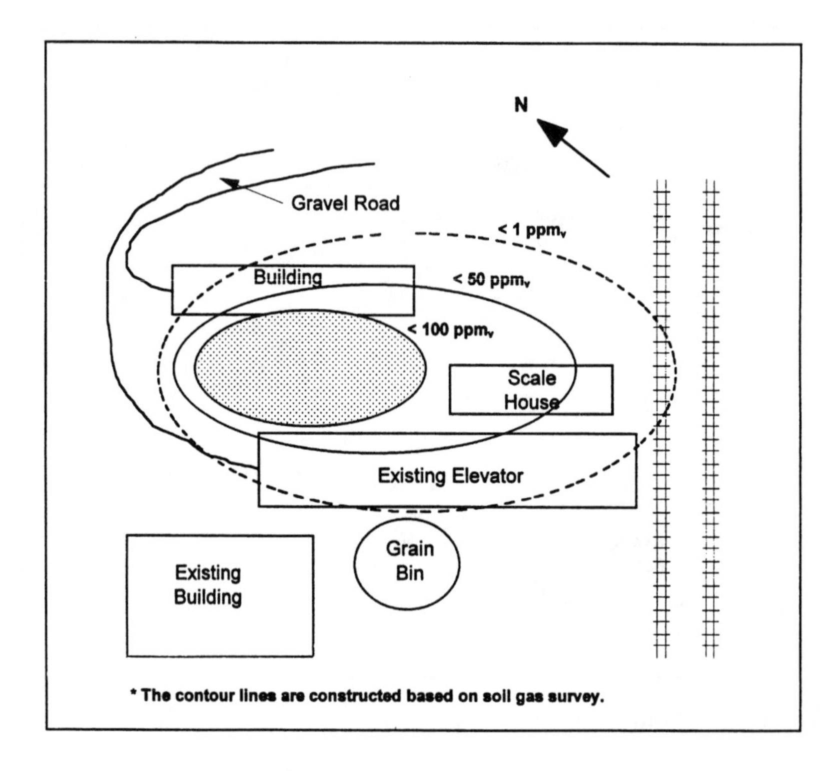

Figure 5.17. Extent of contamination in the vadose zone.

applied to the shallow SVE well was 50, 59, 80, and 100 in. WC. The corresponding air flows at the extraction well were 27, 29, 61, and 74 ACFM, respectively. The ROI at this zone as measured by the adjacent vadose zone monitoring points was approximately 40 to 50 ft.

Carbon tetrachloride concentrations measured at the end of the pilot test were approximately 2,000 ppm$_v$. During the pilot test, approximately 100 gallons of carbon tetrachloride were recovered.

2. Pilot Test in the Deep Zone. The deep SVE well is screened from 60 to 110 ft below bgs in the deep sand zone. The vacuum applied to the deep well were 5, 8, and 12.4 in. WC. The corresponding air flows at the extraction well were 78, 114, and 156 ACFM, respectively. The ROI in this zone as measured by the adjacent vadose zone monitoring points was approximately 150 to 180 ft. The permeability for air in the deep zone was calculated using HyperVentilate to be 10 to 100 Darcy.

Carbon dioxide in the soil vapors extracted at this zone were approximately 6.5% near the end of the pilot test. This is much higher than would be under normal soil conditions (<1%) and implied that biological activity is active in the test zone at this site. Carbon tetrachloride concentrations measured during the test ranged from approximately 5,000 ppm_v at the beginning to 3,000 ppm_v at the end. An estimated 90 gallons of carbon tetrachloride were recovered during the pilot test.

5.7.2.4 Design Objective

1. Removal of carbon tetrachloride in the vadose zone via SVE.
2. Control the further migration of carbon tetrachloride into the groundwater.

5.7.2.5 Design Parameters

1. For the shallow-zone SVE system, a ROI of 40 ft at an applied vacuum of 60 in. WC with 41 ACFM per SVE well will be used.
2. For the deep-zone SVE system, a ROI of 150 ft at an applied vacuum of 12 in. WC with 150 ACFM per SVE well will be used.
3. Off-gas treatment will be accomplished by catalytic oxidation.
4. Other consideration(s): An SVE system specifically designed to target the medial sandy clay zone was originally planned, based on the vacuum responses in the monitoring points screened in 40 to 60 ft bsg during both of the pilot tests. However, the plan was abandoned because the SVE systems in the shallow and deep zones will be able to remediate the contamination in this zone.

5.7.2.6 Conceptual Design

The proposed shallow SVE system will utilize eight vapor extraction wells to remediate the shallow zone. All the SVE wells are screened from 10 to 40 ft bsg and are positioned such that they would cover the area of highest soil contaminations. The combined air flows from eight shallow extraction wells will be approximately 330 ACFM.

The proposed deep SVE system will utilize three vapor extraction wells to remediate the deep zone. All the SVE wells are screened from 60 to 100 ft bsg and are positioned such that they would cover the entire contaminant plume. The combined air flows from three deep extraction wells will be approximately 450 ACFM. Figure 5.18 shows the final design of the SVE systems.

The discharge from both of the SVE systems is then piped to a catalytic oxidation unit for vapor destruction. The combined air flows for both of the systems will be approximately 780 ACFM. A 1,000 ACFM catalytic oxidation unit will be used as the air treatment device. The hydrochloric acid produced by the catalytic oxidation unit will be neutralized with sodium hydroxide and discharged to a nearby creek.

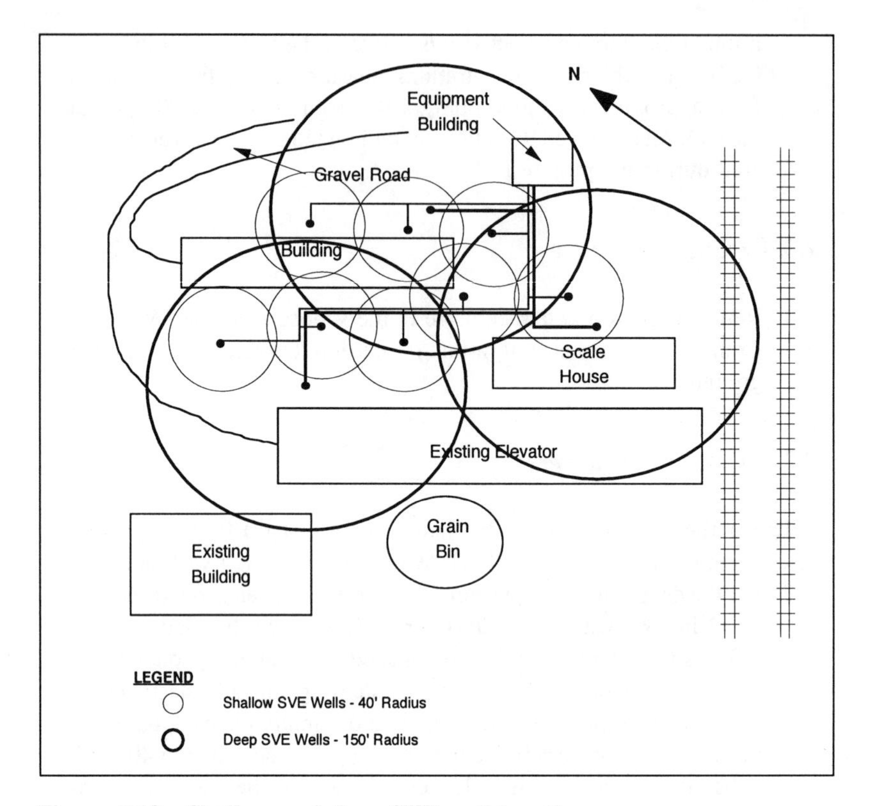

Figure 5.18. Shallow and deep SVE well locations.

6 DESIGN OF BIOREMEDIATION SYSTEMS

6.1 INTRODUCTION

Bioremediation, as inferred by its name, refers to the treatment or remediation of contaminated soils and groundwater using biological means. Municipal wastewater treatment plants have been employing this technology for decades and bioremediation is merely an application of the same principles in a different setting. In some cases, in situ bioremediation occurs as a by-product of another active ongoing remediation system, such as soil vapor extraction (SVE)/air sparging (AS). It should be noted that there is some degree of biological activity occurring in the subsurface at all times and, over time, nature usually heals herself. Bioremediation merely speeds up the process by increasing the rate of bacterial metabolism and growth.

Bioremediation is sometimes referred to by other names such as biorestoration, biodegradation, or bioventing. Bioventing actually refers to the process of supplying air to the vadose zone to stimulate in situ biological activity. Bioremediation treatment systems can be broadly grouped into two categories: in situ and non-in situ. Basically, in situ systems are those where the contaminated medium (soil and/or groundwater) is not physically moved or transported from its original location. Non-in situ, or above-ground, systems involve bringing the contaminated medium to the surface for treatment. Some treatment systems incorporate both in situ and non-in situ aspects in their design.

In situ systems can be further classified as either intrinsic remediation or engineered bioremediation systems (Figure 6.1). Intrinsic bioremediation, also known as natural or passive bioremediation, is essentially allowing nature to take its own course. It should be emphasized, however, that it does not entail a "no action" approach. There is a strict protocol to adhere to when considering intrinsic bioremediation (see section on intrinsic bioremediation). Presence of indigenous bacteria capable of breaking down site-specific contaminants must first be verified. Extensive data collection and plume modeling must be conducted to demonstrate the effectiveness of natural attenuation on a site-by-site basis. This is followed by an extensive long-term monitoring program.

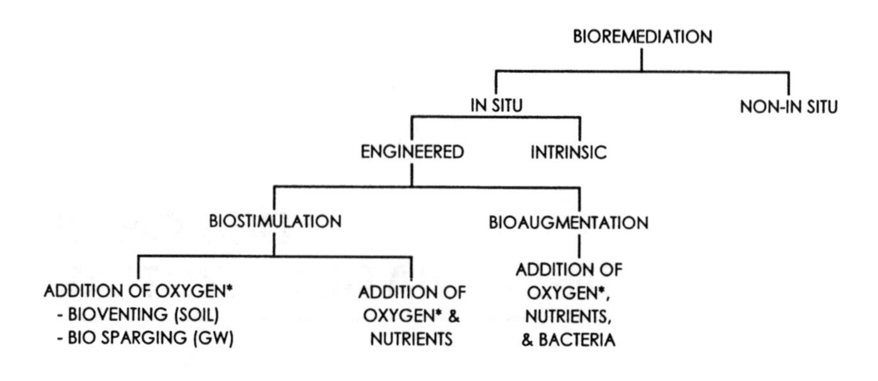

Figure 6.1. Types of bioremediation.

Engineered in situ bioremediation involves the design and installation of systems designed for the purpose of supplying microbe-simulating materials into the subsurface. Engineered systems can, in turn, be broadly categorized as either biostimulation or bioaugmentation systems. Biostimulation refers to the addition of oxygen alone (in aerobic systems) or the addition of both oxygen and nutrients to the subsurface. Bioventing pertains to the process of supplying air to the vadose zone through the process of forced-air injection or vacuum-induced air flow, whereas air sparging or biosparging refers to the process of injecting air into the groundwater under low pressure. Bioaugmentation is the process of adding nonnative bacteria to the subsurface to work together with the indigenous bacteria in breaking down the contaminants. Bioaugmentation typically also includes adding electron acceptors and nutrients.

Non-in situ bioremediation usually involves the design and construction of an above-ground bioreactor or biofilter for the treatment of contaminated groundwater. Biological treatment of excavated soil is also considered a non-in situ method. Landfarming of contaminated soils falls under this classification.

The first commercial in situ bioremediation application was at the Sun Oil pipeline spill site in Ambler, Pennsylvania in 1972 (National Research Council, 1993). Since then, bioremediation has proven to be a viable remedial technology, especially for the treatment of petroleum hydrocarbons. However, when it comes to the treatment of less easily degraded compounds, such as chlorinated solvents, the technology is still considered to be at a research stage.

Overall, bioremediation's role in remediation engineering today is still not quite as established and widely accepted as other technologies such as pump and treat and SVE. However, given the extensive research and development programs undertaken to advance the understanding and application of bioremediation by both public and private sectors, as evidenced by EPA's support and the availability of commercial biosystems, it will probably just be a question of time before bioremediation gains wide recognition as an efficient, cost-effective, and nondisruptive technology.

As non-in situ bioremediation is basically a pump and treat technology, this chapter will deal mainly with the design of in situ treatment systems. Attention will be focused on the design of engineered aerobic systems. Additionally, this chapter will address only the design of treatment systems for remediating volatile organic compounds (VOC), particularly the lighter hydrocarbons, such as gasoline.

6.2 MICROBIOLOGY OF BIOREMEDIATION

The overall concept of bioremediation of petroleum-contaminated sites is fairly simple. Carbon-consuming bacteria occurs naturally in the ground. On these properties, the indigenous bacteria population utilizes petroleum hydrocarbons as a source of food and energy, breaking them down to carbon dioxide and water and producing more biomass. Along with carbon, the bacteria requires an electron acceptor such as oxygen and nutrients such as nitrogen and phosphorus. Bioremediation speeds up the biodegradation process by increasing the electron acceptor and nutrient supplies. When the "nonnatural" food supply (petroleum hydrocarbons) is gone, the majority of the bacteria die.

The detailed chemistry of the hydrocarbon degradation process is far more complex. It entails at least a general understanding of microbiology and the biochemical reactions that accompany the biodegradation process.

The actual application of bioremediation at a site requires knowledge of basic microbiology/biochemical processes, site conditions, and design factors and considerations. The purpose of this section is to advance a basic understanding of the bioremediation process and discuss the conditions and parameters that are favorable to its application. The sections following these will discuss feasibility testing and design aspects of bioremediation systems.

6.2.1 Bacteria

Bacteria are single-celled organisms that metabolize soluble food and reproduce primarily by binary fusion. Bacteria cells generally have one of three shapes: bacillus (rodlike), coccus (spherical or ovoid), and spirillum (spiral or corkscrew). Since microbial growth is the objective of bioremediation systems, its measurement is an important parameter in the design and monitoring of bioremediation systems. There are a number of ways bacterial growth can be measured. The two most commonly used methods are the standard plate count and the most probable number (MPN) method. Both procedures are applicable to soil samples and water samples.

The bacterial cell is made up of approximately 70 to 90% water and 10 to 30% dry mass by weight. Of the dry weight, approximately 92% is composed of carbon, oxygen, nitrogen, and hydrogen. Carbon is the major element, making up between 45 and 55% of the dry cell material. Based on the composition of the cell mass, the most widely accepted empirical formula for the bacterial cell is $C_5H_7O_2N$ (Water Environmental Federation, 1994).

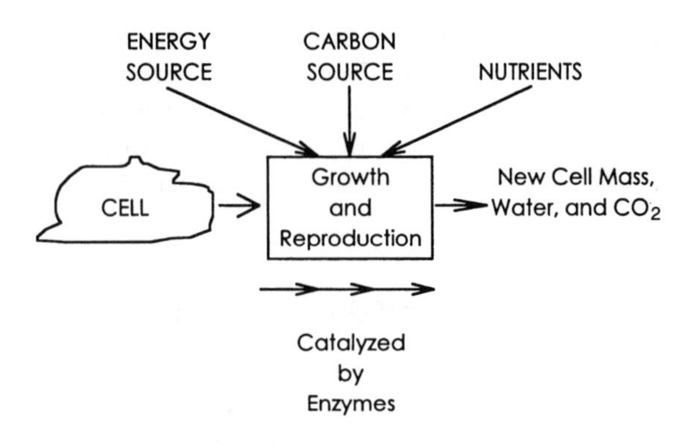

Figure 6.2. Cell metabolism process.

6.2.2 Bacteria Requirements

The requirements for microbial growth may be divided into two categories: physical and chemical. Physical conditions include pH and temperature. Chemically, the bacterial cell needs an energy source, a carbon source, and nutrients from the external environment in order to survive and multiply.

There exist two sources of carbon that bacteria can utilize: organic compounds and carbon dioxide. There are two types of energy sources: chemical compounds or substrates and sunlight. Chemical compounds can be further divided into organic and inorganic sources. Microbes obtain energy from chemicals through the oxidation-reduction process or redox reaction, extracting the energy present in molecules through the transfer of electrons. The organisms use this energy to synthesize new cells and to maintain the old cells already formed.

The energy transfer is not 100% efficient as only a portion of the energy released becomes available for cell use. Nutrients essential for the metabolic process include nitrogen, phosphorus, sulfur, potassium, and iron. The end products of metabolism are water, carbon dioxide, and new cell mass. A simple graphical representation of the basic metabolism process is depicted in Figure 6.2.

In most bioremediation systems, the source of both carbon and energy is the contaminant itself. Nutrients may or may not need to be added to an in situ system, depending on their natural availability. Typically, only the major nutrient requirements, i.e., nitrogen and phosphorus, need to be added. Micronutrients are usually found in sufficient quantities in the soil as trace elements. Most feasibility studies conducted in support of bioremediation do not analyze for micronutrients.

In addition to the above three essential sources that sustain growth and reproduction, two other conditions must exist. First, an electron donor must be present to act as the source of reducing power. Second, there must exist an electron acceptor to oxidize the reducing agent to provide the means of releasing the energy stored in the molecules. In addition, water must also be present as it is an essential component of the metabolic process.

6.2.3 Metabolism

Metabolism is a term used to describe all of the biochemical transformations that occur in living organisms. Metabolic processes provide the mechanism by which cells derive energy and manufacture new cell mass. These metabolic reactions are speeded up or catalyzed by enzymes. Enzymes are protein molecules that act as catalysts to enable biochemical reactions in living organisms to occur at increased rates. The enzyme itself is not changed or consumed in the reaction and is free to take part in additional reactions.

During metabolism, several types of processes take place, including energy conservation, biosynthesis, assimilation and ingestion, and cell maintenance. The complete metabolism process is far more complex than will be covered here. It includes biochemical reactions involving the breakdown of hydrogen atoms, the manufacture of high energy phosphate bonds, the formation of adenosine triphosphate (ATP), molecular diffusion, etc. However, certain fundamental metabolic processes are based on basic chemical reactions and will be discussed here.

These processes are common to all living organisms and are useful for the bioremediation engineer to know. First, the cell uses the carbon from its external carbon source (organic compounds or carbon dioxide) to manufacture cell material (dry weight of cell mass is approximately 50% carbon). Second, there must be a reducing agent (electron donor) present and, third, there must exist an oxidation potential (an electron acceptor or electron "sink" must be present). Cells that feed on organic compounds utilize the hydrogen from the compound itself as their source of reducing power whereas organisms that depend on carbon dioxide for their carbon source must reduce the carbon dioxide to biomass.

In aerobic systems, oxygen is utilized as the electron acceptor. In anaerobic systems, inorganic chemicals such as nitrates, manganese oxides, iron hydroxides, and sulfates are used as substitute electron acceptors. In addition to new cell matter, the by-products of anaerobic respiration may include nitrogen gas, hydrogen sulfide, reduced forms of metals (such as Fe^{2+} and Mn^{4+}), and methane, depending on the electron acceptor.

Before proceeding further, it will be useful to review the oxidation-reduction process or redox reaction, because this is the fundamental way cells obtain the energy present in molocules. Oxidation and reduction reactions involve the transfer of electrons and always go hand-in-hand. Figure 6.3 offers a pictorial illustration of the reaction.

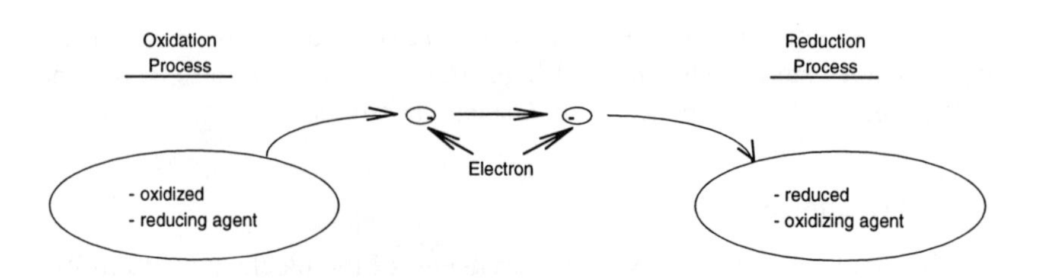

Figure 6.3. Oxidation-reduction reaction.

Oxidation refers to a substance losing electrons and becoming less negatively charged; the substance is said to be oxidized and is called a reducing agent or reducer because it causes the other substance to be reduced. Reduction pertains to a substance gaining electrons and becoming more negative; the substance is thus reduced and is considered an oxidizing agent or oxidant because it causes oxidation. Thus, an oxidant is a substance that causes oxidation to occur while being reduced itself.

The oxidation-reduction process in metabolism is illustrated in Figure 6.4. The electron donor or organic compound is oxidized or gives up hydrogen atoms (electrons) while the electron acceptor is reduced. This is the fundamental way in which cells transform and consume the energy present in molecules to perform work. All cellular production activities and energy conservation mechanisms function in this manner.

In the reaction shown in Figure 6.2, aerobic bacteria convert the organic compound (hydrocarbon contaminants) to carbon dioxide by transferring electrons from the contaminant (oxidizing it) to oxygen (electron acceptor) and reducing it to water. Because of this reaction, a decrease in oxygen levels and an increase in carbon dioxide in the subsurface is usually an indication of microbial activity. This description of the biodegradation pathway is still greatly simplified. Between the original and final products, a succession of new substances has been formed, called intermediates.

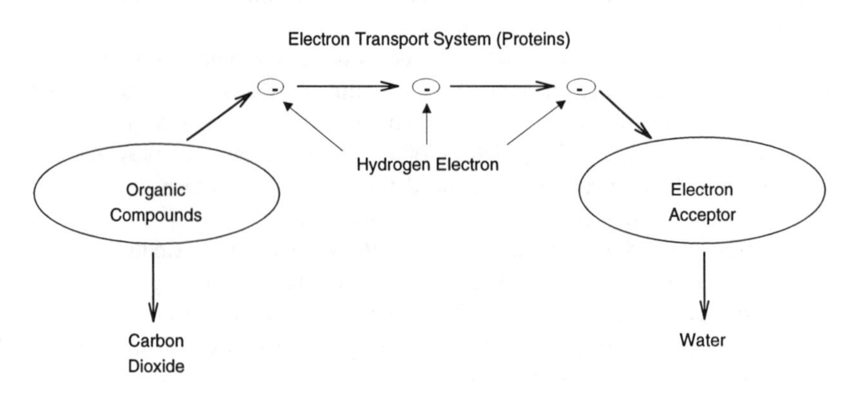

Figure 6.4. Oxidation-reduction in metabolism.

6.2.4 Classification of Bacteria

Bacteria are classified in several different ways, all of which are based on different aspects of the metabolic process. In terms of carbon sources, bacteria are divided into autotrophs, which use carbon dioxide as their carbon source, and heterotrophs, which derive carbon from organic compounds. Additionally, bacteria are grouped according to their source of external energy: phototrophs, which derive energy from photosynthesis, and chemotrophs, which obtain energy from the oxidation of chemical substances.

Based on these two classifications, there can be four basic types of bacteria based on their external sources of energy and carbon. Table 6.1 summarizes these groups in tabular form and describes the major microorganisms in each category. Bacteria that play a key role in bioremediation derive their carbon and energy from organic compounds.

In the wastewater industry, bacteria have traditionally been classified based on their oxygen requirements: aerobic bacteria, which convert their food source (organic compounds) to energy by transferring electrons from the compounds to oxygen (electron acceptor); anaerobic bacteria, which metabolize their food in the absence of oxygen and instead utilize inorganic chemicals such as nitrates, sulfates, carbon dioxide, or metals such as iron as substitute electron acceptors; and facultative bacteria, which can function in both aerobic and anaerobic environments.

In a report published by the National Research Council (NRC), three metabolic processes were identified as playing the most significant roles in bioremediation: aerobic, anaerobic, and cometabolism (National Research Council, 1993). In cometabolism reactions, transformation of a specific contaminant occurs indirectly, i.e., as the result of the metabolism of another substance. For example, certain aerobic bacteria in the process of oxidizing methane (methane serving as the electron donor or the food supply for the microbes and oxygen serving as the electron acceptor), releases certain enzymes that degrade chlorinated solvents, which otherwise could not be metabolized by the bacteria themselves.

6.3 FACTORS AND CONDITIONS AFFECTING BIOREMEDIATION

In situ bioremediation is highly dependent on site conditions and soil properties, more so than SVE/AS. Factors that play a significant role in the design and successful operation of a bioremediation system include contaminant characteristics, natural supplies of macronutrients and micronutrients, availability of electron acceptors, presence of indigenous bacteria capable of degrading the contaminants, and subsurface characteristics.

6.3.1 Contaminant Characteristics

The biodegradability of petroleum products is dependent on the chemical structure of its various components. In general, the lighter, more soluble

Table 6.1 Classifications of Bacteria

	Energy Source		
	Chemicals or Substances (Chemotrophs)		**Light (Phototrophs)**
Carbon Source	**Organic Compounds (Chemoorganotrophs)**	**Inorganic Compounds (Chemolithotrophs)**	
Organic Compounds (Heterotrophs)	Chemoheterotrophs - most bacteria - fungi - protozoa	—	Photoheterotrophs - few bacteria species
Carbon Dioxide (Autotrophs)	—	Chemoautotrophs - some bacteria	Photoautotrophs - algae - some bacteria

Table 6.2 Solubility and Viscosity Data of Representative Petroleum Products

Product	Solubility in Cold Water (at 20 C in ppm)	Viscosity (in Centistokes)
Gasoline	50–100	0.5–0.6
1-Pentene	150	n/a
Benzene	1,791	0.5
Toluene	515	0.5
Ethylbenzene	75	0.6
Xylenes	150	0.6
n-Hexane	12	0.4
Cyclohexane	210	n/a
i-Octane	0.008	n/a
JP-4 Jet Fuel	<1	0.8–1.2
Kerosene	<1	1.5–2
Diesel	<1	2–4
Light Fuel Oil #1 and #2	<1	1.4–3.6
Heavy Fuel Oil #4, #5, and #6	<1	5.8–194
Lubricating Oil	<0.001	400–600
Used Oil	<0.001	40–600
Methanol	>100,000	<0.1

From Cole, M. G., *Assessment and Remediation of Petroleum Contaminated Sites*, CRC Press, Boca Raton, FL, 1994, 63. With permission.

petroleum hydrocarbons are more biodegradable than the heavier, less soluble members of the group. A compound's resistance to biodegradation increases with increasing molecular weight. Additionally, highly viscous hydrocarbons are less successfully biodegraded because of the inherent physical difficulty in establishing contact among the contaminant and the microorganisms, nutrients, and electron acceptors. For example, gasoline, which is considered more easily biodegradable than diesel fuel, has a solubility of 50 to 100 ppm and a viscosity of 0.5 to 0.6 centistokes as compared to a solubility of less than 1 ppm and a viscosity of 2 to 4 centistokes for diesel (Table 6.2) (Cole, 1994). Similarly, diesel is more biodegradable than used oil, which has a solubility of less than 1 ppb and a viscosity of 40 to 600 centistokes.

Simpler chemical structures are also easier to degrade. Branched structures degrade at a slower rate than the corresponding straight-chain hydrocarbons. Alkanes are degraded more rapidly than aromatic compounds. Monoaromatic compounds such as BTEX are broken down faster than the two-ring

Table 6.3 Examples of Organic Compounds That Have Been Shown to Be Biodegradable

Acenaphthalene	Fluoranthene
Acenamphthene	Heptane
Acetone	Hexane
Antracene	Isopropyl Acetate
Benzene	Methylene Chloride
Benzo(a)anthracene	Methyl Ethyl Ketone
Benzo(k)fluoranthene	Methylmethacrylate
Benzo(a)pyrene	Naphthalene
Butanol	Nitroglycerine
Carbon Tetrachloride	Nonane
Chlorobenzene	Oil & Grease
Chloroethane (CA)	Octane
Chlorophenols	Pentachlorophenol (PCP)
Chloroform	Perchloroethylene (PCE)
Chrysene	Phenanthrene
p-Cresol	Phenol
Dibenzo(a,h)anthracene	Phytane
DDT	Polychlorinated Biphenyls
Dichlorobenzene	Pristane
Dichloroethane (DCA)	Pyrene
Dichloroethylene (DCE)	Styrene
Dichloromethane (DCM)	Tetrachloroethylene (PCE)
Dioxane	Toluene
Dioxin	1,1,1-Trichloroethane (1,1,1-TCA)
Dodecane	Trichloroethylene
Ethylbenzene	Tridecene
Ethyl Glycol	Vinyl Chloride (VC)
	Xylene

compounds such as naphthalene. Table 6.3 provides examples of organic compounds that have been shown to be biodegradable.

Some chemicals may be toxic to the microbes. In some cases, compounds that are readily biodegradable in low concentratons may exhibit toxicity characteristics to the microorganisms at high levels. Lead present in unleaded gasoline may also contribute to the toxicity of gasoline to the bacteria.

6.3.2 Indigenous Microorganisms

The microorganisms generally attributed for the bioremediation of contaminants are generally bacteria, but indigenous fungi are also responsible for hydrocarbon degradation. For complete degradation of the lighter hydrocarbons, multiple strains of bacteria are required. The indigenous bacteria

Table 6.4 Genera of Hydrocarbon-Degrading Bacteria and Fungi Isolated from Soil

Bacteria	Fungi
Achromobacter	*Acremonium*
Acinetobacter	*Aspergillus*
Alcaligenes	*Aureobasidium*
Arthrobacter	*Beauveria*
Bacillus	*Botrytis*
Brevibacterium	*Candida*
Chromobacterium	*Chrysosporium*
Corynebacterium	*Cladosporium*
Cytophaga	*Cochliobolus*
Erwinia	*Cylindrocarpon*
Flavobacterium	*Debaryomyces*
Micrococcus	*Fusarium*
Mycobacterium	*Geotrichum*
Nocardia	*Gliocladium*
Proteus	*Graphium*
Pseudomonas	*Humicola*
Sarcina	*Monilia*
Serratia	*Mortierella*
Spirillum	*Paecilomyces*
Streptomyces	*Penicillium*
Vibrio	*Phorma*
Xanthomonas	*Rhodotorula*
	Saccharomyces
	Scolecobasidium
	Sporobolomyces
	Sprotrichum
	Spicaria
	Tolypocladium
	Torulopsis
	Trichoderma
	Verticillium

From Calabrese, E. J. and P. T. Kostecki, *Principles and Practices for Petroleum*, CRC Lewis Publishers, Boca Raton, FL, 1993. With permission.

population in the soil generally contains the necessary mixtures of bacteria species to perform that function. Table 6.4 lists, in alphabetical order, the genera of hydrocarbon-consuming bacteria isolated from soil. The most commonly isolated species are *Pseudomonas*, *Arthrobacter*, *Alcaligenes*, *Corynebacterium*, *Flavobacterium*, *Achromobacter*, *Micrococcus*, *Nocardia*, and *Mycobacterium*.

For the heavier hydrocarbons, commercial preparations of bacteria may need to be added to supplement the native microbial population.

Petroleum-degrading bacteria are not the only microbes of interest in bioremediation. Iron bacteria, that notorious group of aerobic bacteria that oxidize ferrous (dissolved) iron as a source of energy, can cause serious operational problems for the treatment system.

6.3.3 Nutrients

In addition to carbon, there are other important elements required by microorganisms. Based on the amounts required in the metabolic process, these elements are classified as macronutrients and micronutrients.

6.3.3.1 Macronutrients

The macronutrients required by the bacteria population are carbon, nitrogen, and phosphorus. The optimum carbon:nitrogen:phosphorus (C:N:P) ratio is 100:10:1. The carbon supply comes from the hydrocarbon contaminants themselves. Nitrogen and phosphorus may be naturally present in the ground in sufficient amounts. Treatability pilot studies designed for assessing the feasibility of in situ bioremediation typically analyze for nitrogen in the form of ammonia and nitrate, and phosphorus in the form of orthophosphate.

Ammonium and nitrate salts are the most common sources of nitrogen, whereas orthophosphate and tripolyphosphate salts are the most widely used sources of phosphorus (National Research Council, 1993). Numerous brands of biological products for enhancing bacterial growth and activity are available commercially. These products are usually mixtures of nitrogen, phosphorus, trace elements, enzymes, and buffering and neutralizing compounds. Pure sources of nitrogen and/or phosphorus are also available.

6.3.3.2 Micronutrients

Micronutrients that must be present in the soil include sulfur, potassium, sodium, calcium, magnesium, iron, manganese, zinc, and copper. Additional trace elements required for anaerobic metabolism include cobalt and nickel. Many commercially available fertilizers contain adequate quantities of trace elements. For optimum sustained microbial growth and reproduction, both macro- and micronutrients must be present, and must be in the proper amounts, forms, and ratios.

6.3.4 Electron Acceptors

In aerobic metabolism, molecular oxygen (O_2) acts as the terminal electron acceptor. Approximately three pounds of available oxygen are required to convert one pound of hydrocarbon to carbon dioxide and water. Using toluene (C_7H_8) as an example to illustrate stoichiometrically the breakdown of petroleum hydrocarbons aerobically, the chemical equation will be as follows:

$$C_7H_8 + 9O_2 \rightarrow 7CO_2 + 4H_2O$$

In anaerobic bioremediation, alternate or substitute electron acceptors are used in place of oxygen. These include, in order of preference, nitrate (NO_3^-), manganese (IV) and iron (III) oxides (MnO_2 and $Fe(OH)_3$, respectively), sulfate (SO_4^{2-}), and carbon dioxide (methanogenesis). In general, the use of a particular electron acceptor is a function of its availability, the presence of other electron acceptors, and the related oxidation-reduction potential of the surrounding environment. The sequence of microbially mediated redox processes as a function of the redox potential of the groundwater is depicted in Figure 6.5. The energy yield for microorganisms from hydrocarbon metabolism varies greatly. Oxygen is preferred because microorganisms gain the most energy from aerobic reactions. The energy yield is generally in the following order:

$$O_2 > NO_3^- > Mn^{4+} > Fe^{2+} > SO_4^{2-} > CO_2$$

Again, using toluene as the representative chemical to be biodegraded, the stoichiometric formula for the breakdown with the substitute electron acceptors are as follows:

Nitrate Reduction:

$$C_7H_8 + 6NO_3^- \rightarrow 7CO_2 + 4H_2O + 3N_2$$

There exist certain situations when nitrate may be more adequately utilized as an electron acceptor than oxygen. First, nitrate is more soluble than oxygen and, second, it is less expensive. Thus, from a practical standpoint, it may be more economical to use nitrate rather than oxygen. Additionally, aerobic remediation of deeper zones of the aquifer may not be feasible. On the negative side, nitrate is a listed pollutant under the National Primary Drinking Water Regulations with a maximum contaminant level (MCL) of 10 mg/L and its introduction into the aquifer may face regulatory scrutiny.

Manganese Reduction:

$$C_7H_8 + 9MnO_2 \rightarrow 9Mn^{2+} + 7CO_2 + 4H_2O$$

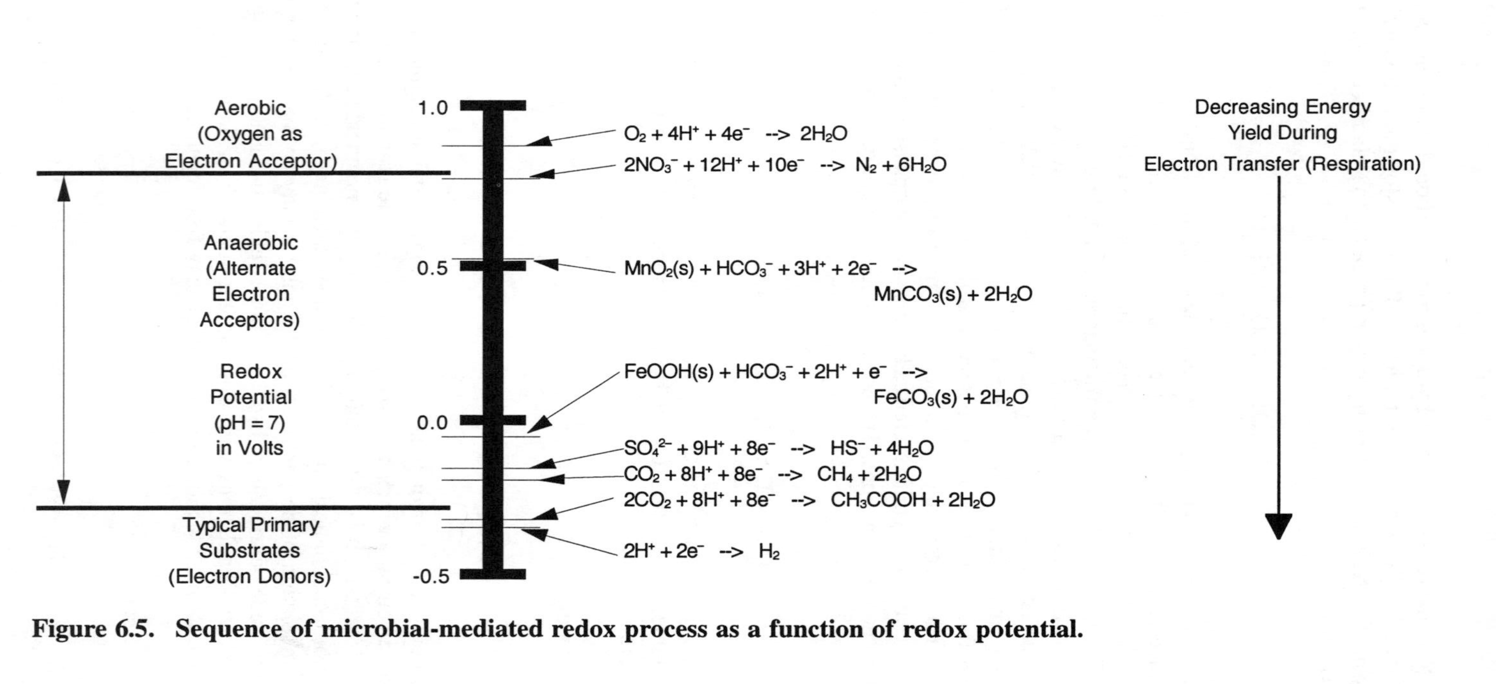

Figure 6.5. Sequence of microbial-mediated redox process as a function of redox potential.

Insoluble manganese oxide, MnO_2 is reduced to dissolved manganese (II) by hydrocarbon-degrading bacteria. The level of manganese in groundwater seldom exceeds 2 mg/L. Where anaerobic conditions have developed at a petroleum-contaminated site, it is likely that concentrations of manganese (II) will be higher.

Iron Reduction:

$$C_7H_8 + 36Fe(OH)_3 \rightarrow 7CO_2 + 36Fe^{2+} + 72OH^- + 22H_2O$$

Iron is found in the subsurface as either dissolved ferrous, reduced iron (II), or insoluble ferric, oxidized iron (III). In obtaining energy for their metabolic needs, iron bacteria such as ferrobacillus, gallionelle, and sphaerotilus oxidize ferrous iron to ferric iron. Significant amounts of precipitated ferric iron may be present in the aquifer, especially under aerobic conditions.

In the biodegradation of hydrocarbons, carbon-consuming bacteria will, in the process of oxidizing the organic compounds, reduce ferric iron to soluble ferrous iron. Field results have indicated an increase in dissolved ferrous iron accompanied by a corresponding decrease in hydrated iron (III) in an aquifer contaminated with crude oil (Lovley et al., 1989). Under normal conditions, dissolved iron levels in groundwater seldom exceed 10 mg/L. The reduction of iron (III) in hydrocarbon-contaminated aquifers could result in concentrations of dissolved iron (II) at levels as high as 10 to 100 mg/L.

Sulfate Reduction:

$$4C_7H_8 + 18SO_4^{2-} + 12H_2O \rightarrow 9H_2S^- + 9HS + 28HCO_3^- + H^+$$

After dissolved oxygen, nitrate, manganese, and iron have been depleted, sulfate may be used as an electron acceptor for biodegradation. Sulfate reduction is the anaerobic microbial reduction of the electron acceptor, sulfate to hydrogen sulfide gas. This process is termed sulfanogenesis and it precedes methanogenesis or methane formation, described below. Presently, little is known about the significance of hydrocarbon biodegradation through the use of sulfates as the electron acceptor.

Methanogenesis:

$$2C_7H_8 + 10H_2O \rightarrow 9CH_4 + 5CO_2$$

Methanogenesis is the conversion of low molecular fatty acids, alcohols, carbon dioxide, and hydrogen to methane gas. It follows sulfate reduction. Both processes are accomplished by a group of bacteria known as methanogenic bacteria. While methanogenesis is a familiar process in traditional wastewater engineering (anaerobic digestion of sludge), its significance in hydrocarbon biodegradation is poorly understood. Vogel and Grbic-Galic described the potential reaction with the above theoretical equation.

6.3.5 Subsurface Characteristics

Of all the factors and conditions affecting bioremediation, subsurface characteristics may be the most critical. This is because the design engineer has little control over the conditions he or she has to work with beneath the ground surface. Where there is a deficiency of electron acceptors or nutrients, delivery systems can be designed to provide them. Even where a certain species of bacteria required for the biodegradation process is absent, the bacteria can be cultured in the laboratory and introduced into the subsurface (bioaugmentation). However, little can be done to compensate for site subsurface conditions that simply do not promote successful bioremediation or are far from ideal for implementing biorestoration systems.

There are many soil properties and groundwater parameters that influence the bioremediation process. These include soil type, soil permeability, grain size distribution, soil moisture content, pH, temperature, groundwater geochemistry, groundwater conductivity, and depth to the groundwater.

6.3.5.1 Soil Type

Soil type is an important, sometimes overlooked variable in bioremediation designs. In general, noncohesive soils such as gravel and sands are the most favorable for the application of bioremediation, whereas tight materials such as clay are the least favorable. Soil permeability is a key factor in the success of bioremediation. Because it will be easier to transport and distribute nutrients and electron acceptors in the subsurface in permeable soils, it therefore follows that the more permeable the soil, the more favorable the condition it is for bioremediation. This is true for both vadose zone and saturated zone bioremediation.

The introduction and movement of air in soils with high permeability through bioventing is more successful than in soils with low permeability. Likewise, groundwater circulation and distribution in a saturated zone treatment system occurs at a much faster rate in soils with high permeability. This is true for both in situ and above-ground treatment systems.

Besides the ease of transport, permeability also plays an important role in the prevention of excessive aquifer plugging. Because of the nature of the microbiology and geochemistry of groundwater systems, total prevention of plugging and fouling in the water-bearing zone would be impossible. The biodegradation process itself produces more microbial mass. Additionally, minerals in the groundwater will oxidize to insoluble oxides and hydroxides with the introduction of oxygen into the system.

The key is to prevent or avoid extensive biofouling and chemical precipitation to the point where groundwater movement is restricted. Thus, the water-bearing zone should be sufficiently permeable so as to prevent the increased microbial mass from causing excessive plugging of the aquifer pores. Microbial growth can reduce aquifer permeabilities by as much as a factor of 1000 (Taylor et al., 1990). It has been suggested that sites with overall hydraulic conductivities of 10^{-4} cm/s or greater would be suitable for in situ bioremediation (Thomas and Ward, 1989).

Besides permeability, grain size distribution may also play a significant role in the prevention of biofouling. Studies have shown that high-porosity materials that are well-graded, i.e., with widely distributed grain sizes, especially with small-diameter pore size ranges, are much more susceptible to biofouling than high-porosity materials with a limited pore size range (Taylor and Jaffe, 1991). Thus, a free-draining material is likely more amenable to bioremediation than a well-graded material.

6.3.5.2 Soil Moisture Content

In vadose zone treatment systems, adequate moisture in the soil is important because microorganisms need water to support their metabolic process. A soil moisture content of 50% is generally considered ideal for bioremediation. One possible disadvantage to bioventing systems is the decrease in moisture content in the soil over time.

6.3.5.3 Groundwater Chemistry

The geochemistry of the affected water-bearing zone or aquifer plays a significant role in the bioremediation process. The viability of in situ treatment systems are greatly influenced by the presence and concentrations of certain chemical species in the groundwater. Perhaps the most important of these are the concentrations of iron and, to a much lesser extent, manganese.

In most groundwater systems with low dissolved oxygen and high carbon dioxide levels (anaerobic or reducing conditions), the iron and manganese are generally present in their dissolved or reduced states, Fe^{2+} (ferrous iron) and Mn^{2+}. Because Fe^{2+} and Mn^{2+} are thermodynamically stable only in an anaerobic environment, it follows that the development of anaerobic conditions is essential for any significant presence of soluble or dissolved iron and manganese in the water system.

The redox potential, Eh, of the groundwater, measured in mV, is another indicator of aerobic/anaerobic conditions. It is a measure of the state of oxidation or reduction in the system. A high redox potential indicates an aerobic environment (oxidizing conditions), whereas a low redox potential indicates an anaerobic environment (reducing conditions). The redox potential of the groundwater can provide indications of the type of redox reactions taking place in the aquifer. This is because microbially mediated redox reactions follow a specific sequence based on the redox potential of the surroundings. Figure 6.5 illustrates the redox potentials for various electron acceptors.

By introducing oxygen into the vadose zone, whether through mechanical or chemical means, the redox potential of the water bearing zone is increased as dissolved oxygen levels increase. An increase in the dissolved oxygen content in the aquifer can produce several chemical and biological reactions in the aquifer locally. These are summarized below:

- Oxidation of hydrocarbons and other organic material by indigeneous aerobic microorganisms. This results in an increase in cell biomass as the growth in microbe population is facilitated by the introduction of oxygen.
- Oxidation of iron and manganese with molecular oxygen, which, depending on the concentrations of dissolved iron and manganese in the groundwater, will most likely lead to precipitation problems. The rate of oxidation of ferrous iron to ferric hydroxide increases as pH increases. At a pH of 6, the reaction is slow; at a pH of 7.5, the reaction becomes very rapid (American Water Work Association, 1990). By keeping the pH below 6, the precipitation of iron can be slowed down, thus reduced forms can persist for some time even in aerated waters.
- Oxidation of iron by iron bacteria. Just as petroleum-degrading bacteria obtain energy from the oxidation of hydrocarbons, iron bacteria derive energy from the oxidation of ferrous iron to ferric iron. Troublesome conditions often result because the ferric iron precipitates in the gelatinous sheaths of the microbial deposits, creating a wet, slimy mass of bacterial growth and insoluble iron hydroxides.
- Consumption of free hydrogen ions, thus leading to an increase in pH.

$$1/2\ O_2 + 2H^+ + e = 2H_2$$

Specific conductivity, measured in micromhos, provides a rapid estimate of the dissolved solids content of the groundwater. Most dissolved inorganics in groundwaters exist in the ionized form, and therefore, contribute to electrical conductivity. Thus, as the ion concentration of the groundwater increases, its conductivity also increases. As a rule of thumb, the specific conductance of most groundwaters (in micromhos) multiplied by a factor of 0.55 to 0.75 provides a reasonable estimate of the dissolved solids content (in mg/L) (Driscoll, 1986). A high specific conductance may affect the movement of bacteria in the aquifer. Experimental studies have demonstrated that bacteria movement can be enhanced by reducing the ionic strength of the water (U.S. EPA, 1991a).

The favorable pH range for bioremediation is between 6 and 8, with the ideal pH at around 7 (neutral). At a site where high levels of bacterial activity is occurring, there is a possibility for the soil conditions to turn slightly acidic due to the production of intermediate acidic compounds. Additionally, carbon dioxide, an end product of the metabolism of hydrocarbons, will contribute to the lower pH as dissolved carbon dioxide forms carbonic acid. Most of the bioremediation supplemental products available also provide buffering and neutralizing agents.

An increase in temperature is followed by a corresponding increase in bioactivity. Hydrocarbon biodegradation rates approximately double for every ±10°C increase in temperature over the temperature range between ±5°C and ±25°C. In addition, groundwater temperature can affect the availability of oxygen as the solubility of oxygen is temperature dependent. Oxygen is more soluble in cold water than in warm water. One of the advantages of in situ bioremediation is that the temperature in the subsurface remains fairly constant year round.

6.3.6 Design Considerations

The design of a bioremediation system involves providing a system for moving electron acceptors, nutrients, and microorganisms to the area of contamination. The physical and mechanical layout of bioremediation systems can take on several different configurations. The treatment of contaminated soils in the vadose zone typically consists of bioventing wells for providing oxygen and a series of infiltration galleries and/or injection wells for the delivery of nutrient-enriched moisture, as depicted in Figure 6.6.

Bioventing wells circulate air through the unsaturated zone either by creating a vacuum (similar to SVE) or through the direct injection of air (at low pressure and low flowrates). Oxygen is generally the only electron acceptor introduced in the treatment of contaminated soils in the vadose zone. In cometabolism, methane gas is injected along with air into the subsurface for the remediation of chlorinated hydrocarbons. Nitrous oxide, for the purpose of providing nitrogen, is sometimes injected along with air.

In situ bioremediation systems for the treatment of contaminated groundwater typically consists of a combination of injection wells and infiltration galleries, and one or more recovery wells as shown in Figure 6.7. This system promotes bioactivity in the vadose zone, too. Oxygen is provided by the injection wells. The combined action of injection and extraction wells help provide circulation of dissolved oxygen and nutrients through the contaminated zones.

Besides providing a mechanism for dissolved oxygen and nutrient transport, groundwater pumping also ensures plume control and contaminant capture. An alternative system utilizes a combination of air sparge wells and infiltration galleries in tandem with vapor extraction wells as illustrated in Figure 6.8. Variations of these systems can also be implemented. Biosparging,

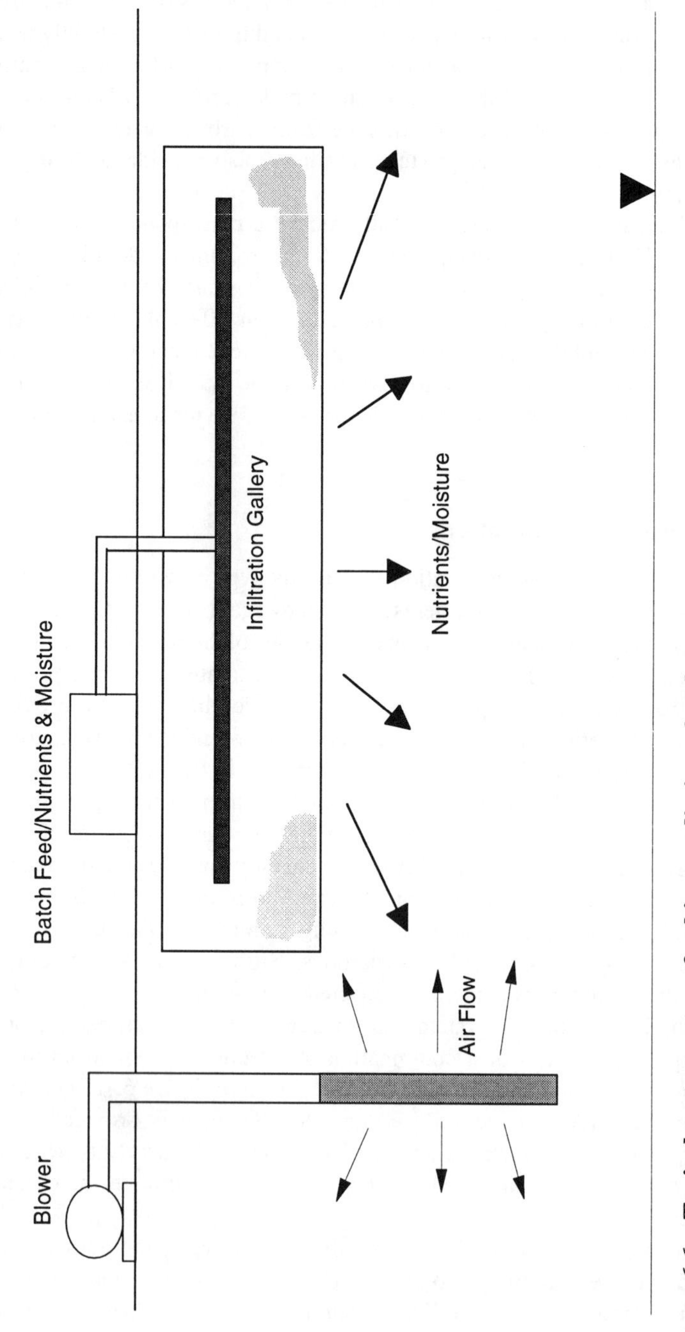

Figure 6.6. Typical system set-up for bioremediation of vadose zone.

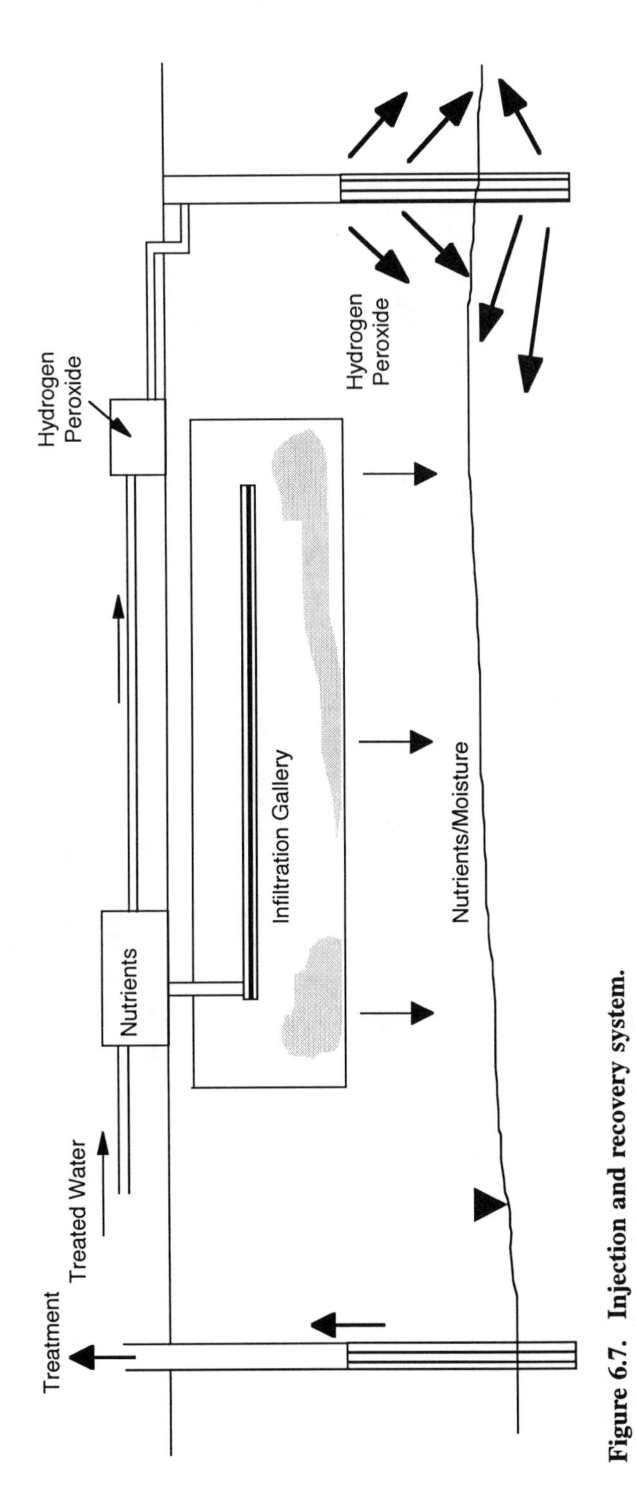

Figure 6.7. Injection and recovery system.

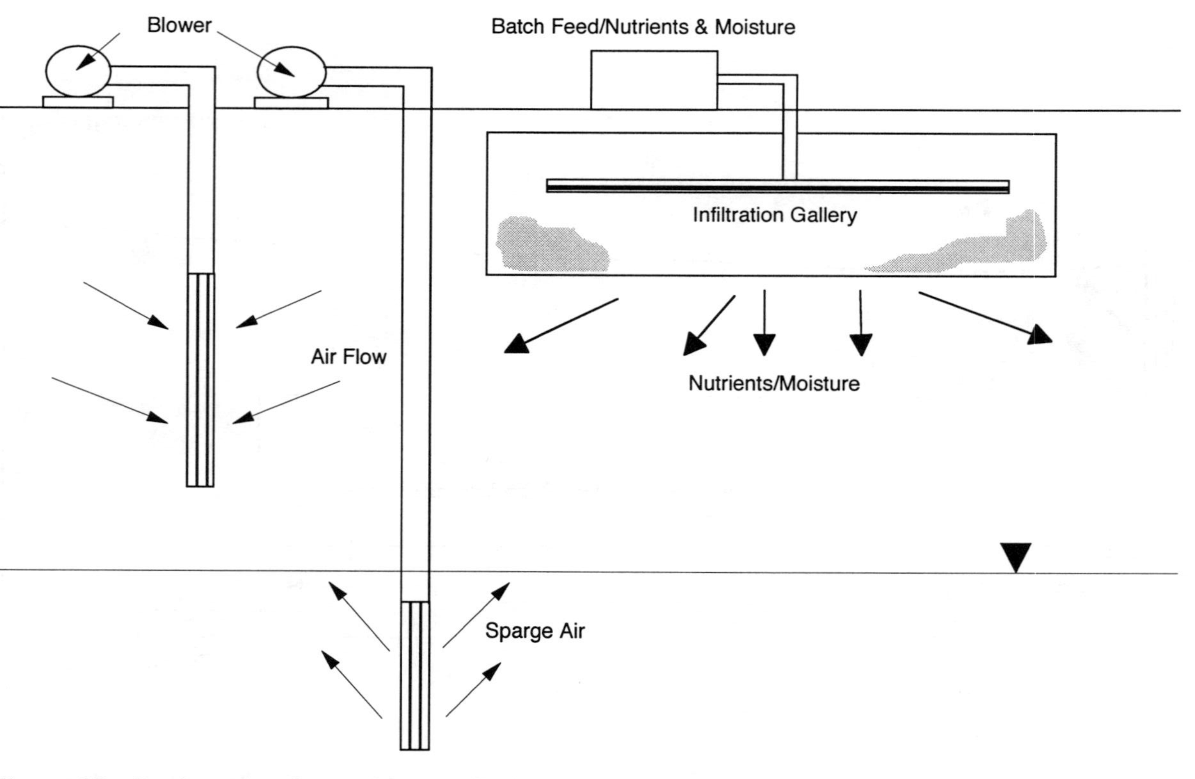

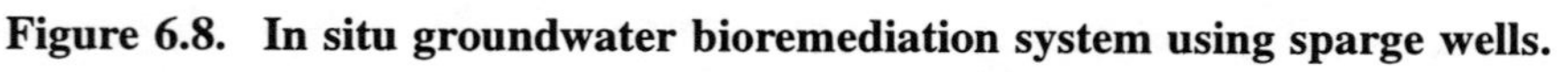

Figure 6.8. In situ groundwater bioremediation system using sparge wells.

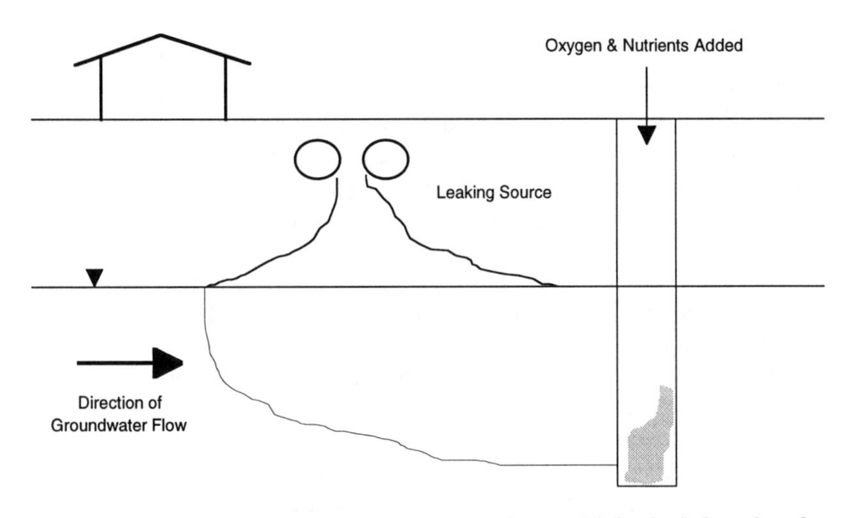

Figure 6.9. Containment trench — creating a biological barrier for degrading contaminants.

injection of air at low pressures as opposed to air sparging, can be deployed instead. In this case, SVE would most likely not be required.

Increasingly, bioremediation is being considered for implementation as a containment system. This is a passive treatment alternative and is attractive at sites where site-specific conditions do not warrant active remediation of the groundwater. A typical biocontainment system consists of a trench or series of wells designed to create a "bio"-barrier or wall (Figure 6.9). Oxygen and nutrients provided to the barrier create an in situ bioreactor that promotes microbial growth and activity.

Depending on site conditions, such as the required depth of the trench (function of depth to groundwater and contaminant type) and chemistry of the groundwater, anaerobic electron acceptors may be utilized. The barrier is usually placed perpendicular to the direction of groundwater flow at a downgradient location. The barrier serves to prevent off-site migration of contaminants.

Essentially, the basic objective of all bioremediation systems is to create a favorable environment for contaminant-degrading microorganisms to flourish in. This is true regardless of whether the environment refers to the fixed-volume of an above-ground bioreactor, a subsurface zone approximately delineated only by the extent of the contamination, or an underground gravel-filled barrier trench. As such, the fundamental aspect in the design of bioremediation systems pertains to the provision of adequate electron acceptors and nutrients to the microorganisms. Obviously, different mechanisms are deployed depending on whether the system is designed for soil or groundwater treatment.

6.3.6.1 Providing Electron Acceptors

In aerobic systems the concern is with providing adequate supplies of oxygen. There are generally two different methods for introducing oxygen into the subsurface: through physical systems and through chemical means. Physical means involve providing air, either in gaseous or liquid phase, through mechanical devices. Chemical methods involve the injection of chemicals, which are then converted into oxygen.

Bioventing is the process of supplying oxygen (air) to the vadose zone either by creating a vacuum and drawing a flux (air flow) through the soil or by injecting air into the soil. The former is similar to SVE; however, the major objective of SVE is volatilization of hydrocarbons, and the objective of bioventing is bioremediation. As such, the vacuum applied is usually kept to a level so as to minimize the volatilization of vapors that would then need to be treated. Bioventing is used only for delivering oxygen to the vadose zone. Some areas of the saturated zone may be affected if used in conjuntion with dewatering wells.

Mechanical means of introducing oxygen into the saturated zone include air sparging and aerated water injection. Sparging is the process of injecting air under pressure into the saturated zone below the groundwater table. This method is similar to that used in SVE/AS systems. Usually the air is injected at a lower pressure so as to prevent the migration of contaminated vapors into the vadose zone.

Air could also be bubbled into the sparge well and allowed to diffuse into the surrounding formation. This is accomplished by using a porous stone or fitted glass diffuser. If higher sparge pressures are applied, the air sparging system will have to operated in tandem with an SVE system. It is important to remember that the objective of biosparging wells is to provide adequate air to increase dissolved oxygen but avoid significant volatilization of contaminants. Pure oxygen can also be used in place of ambient air in sparge wells. This will increase the dissolved oxygen content in the groundwater; however, it will increase the cost of the system.

Aerated water injection involves the injection of oxygen-saturated water into the water-bearing zone. The limiting factor is the solubility of oxygen in water, since water saturated with air at 25°C (77°F) contains approximately 8 to 10 mg/L of oxygen.

The chemical method involves the injection of hydrogen peroxide, H_2O_2, into the subsurface. Hydrogen peroxide is mixed with water and circulated in the subsurface through injection wells that may be screened both in the vadose and saturated zones. Hydrogen peroxide decomposes as follows:

$$2H_2O_2 \rightarrow O_2 + 2H_2O$$

Studies have shown that the usage of hydrogen peroxide has its limitations. Because it will oxidize any reduced substance, hydrogen peroxide has a ten-

Table 6.5 Comparison of Oxygen Availability Between Air Sparging and Hydrogen Peroxide, lb/day

Air Sparging				Hydrogen Peroxide (1,000 ppm)			
Flow	**Utilization**			**Flow**	**Utilization**		
SCFM	100%	50%	10%	GPM	100%	50%	10%
10	236	118	24	10	56	28	6
25	590	295	59	25	140	70	14
50	1182	590	118	50	280	140	28

dency to prematurely decompose. This restricts its availability to the targeted microorganisms. Table 6.5 provides a comparison of the actual amount of oxygen available for bioremediation between air sparging and hydrogen peroxide injection.

Recently, solid materials that gradually release oxygen when in contact with water have been developed. These materials can be particularly effective in groundwater containment systems where they form an in situ biological barrier.

6.3.6.2 Providing Nutrients

For in situ systems in saturated environments, groundwater is typically recovered, treated, amended with nutrients, and introduced back into the subsurface either by injection wells or infiltration galleries. For vadose zone systems, nutrients are mixed with water in a mixing tank and either percolated into the soil through infiltration galleries or injected through injection wells. Nitrous oxide gas can also be injected into the vadose zone with air for the provision nitrogen.

6.3.6.3 Providing Microorganisms

For the in situ treatment of gasoline-contaminated sites, bioaugmentation is seldom a consideration. Petroleum-degrading microorganisms are commonly part of the indigenous microbial population.

6.4 FEASIBILITY STUDIES

Like most long-term treatment methods in the remediation of contaminated sites, a feasibility determination or pilot test of the proposed technology should always be performed prior to detailed design and construction. As with other remedial technologies such as SVE and AS, the pilot test presents data and results that enable the design engineer to evaluate the feasibility of the proposed system, and to design the system. However, unlike other technologies such as SVE, AS, and pump and treat, measuring the success of in situ bioremediation is not as simple and straight-forward.

The feasibility of SVE/AS systems can be determined by examining applied vacuum and flowrate relationships, and ROI achieved. Evidence of pump and treat viability can be determined through aquifer pump test results. In in situ bioremediation, however, microbial cleanup can also be attributed to other factors, such as volatilization, if oxygen is introduced into the subsurface as an electron acceptor, and off-site migration, especially in highly permeable material. Additionally, the rate of contaminant removal or destruction occurs at a slower rate in bioremediation than with volatilization processes.

Measurements of in situ bioremediation success is further compounded by the inherent difficulty in analyzing a process that is taking place beneath the ground surface. Pilot tests for bioremediation systems are conducted for much longer durations than other treatment systems. In general, bioremediation studies can last anywhere from three weeks to six months, depending on the type of contaminant and subsurface conditions. Feasibility studies on sites impacted with petroleum hydrocarbon products such as gasoline typically require less time than those involving chlorinated solvents or heavier hydrocarbons.

An excellent procedure for evaluating the feasibility and effectiveness of in situ bioremediation was compiled by a special committee of the National Research Council (1993). A threefold process is suggested, involving the following steps:

1. Demonstrate a reduction in contaminant concentrations.
2. Prove that the native microorganisms at the site can degrade the contaminants.
3. Evidence that biodegradation is indeed taking place at the site.

The strategy recommended above applies to both the evaluation of feasibility tests and the monitoring of ongoing systems. By meeting the strategy's three requirements, evidence can be presented to demonstrate the success of the system.

The first step is mandatory in almost all remediation projects. Documenting decreasing contaminant levels is accomplished through the sampling and testing of soil and groundwater samples on a regular basis over time. The second requirement, demonstrating the capability of the microbes to breakdown the contaminants can be achieved either through laboratory assays or through literature reviews.

The third phase of the strategy, showing field biodegradation, is the most elaborate and difficult. There exist various techniques or indicators for demonstrating on-site biodegradation. The methods that are typically used are summarized in Table 6.6. Basically, the techniques strive to prove that biodegradation is taking place by showing that the biological and chemical changes occurring on-site are in line with what one would expect when bioremediation is occurring. Typically, several different techniques are utilized at any one site.

Table 6.6 Indicators of On-Site Biodegradation

Indicator	Method	Remarks
Increase in Bacteria Population	Total Heterotropic Bacteria Plate Count	Correlate decrease in contaminant with increase in bacteria population. However, it should be noted that because of natural variations in bacteria populations, it may be difficult to establish a significant trend, especially over a short period of time.
Changes in Electron Acceptor Concentrations	Standard Analytical Methods from Wet Chemistry Field Measurement Meters Field Test Kits	A depletion of electron acceptor concentrations that occur simultaneously with contamination reduction. An increase in electron acceptor as a result of continued electron acceptor delivery, along with contaminant loss.
Increase in By-Products of Biodegradation	Standard Chemical Analytical Methods Field Test Kits	Increase in carbon dioxide and methane gas concentration are indicative of bacterial activity.

6.4.1 Data Collection

The major parameter that influences the feasibility of using bioremediation as a treatment method is the biodegradability of the contaminants of concern. Prior to conducting field sampling and data collection, it should first be confirmed that the targeted compounds are indeed amenable to biological treatment. Table 6.3 provides a partial list of compounds that have been shown to be biodegradable.

In order to achieve the goals of the feasibility test, a baseline database of the site conditions will first have to be established before proceeding with bench-scale in situ remediation. These data will later be used for comparison during the biodegradation process. In general, the data that need to be collected are essentially functions of the conditions that affect the bioremediation process.

- Contaminanant characteristics — this information, including type of contaminants and concentrations, will most likely be available from the site assessment and hydrogeological reports. Contaminant levels prior to initiating remediation is critical for documenting loss of contaminants from the site.
- Indigenous microorganisms — soil and/or groundwater samples should be collected and analyzed for the presence of naturally occur-

ring microorganisms. It should be noted that bacteria plate counts from groundwater samples are usually on the low side and are not indicative of the true bacteria population in the subsurface. This is because the bacteria population tends to be attached to the soil matrix. A bacteria genera identification test should also be performed.

- Macronutrients and micronutrients — soil and/or groundwater samples should be collected for nutrient optimization studies. Typically, only the macronutrients, nitrogen, and phosphorus are analyzed. Nitrogen is generally analyzed in the nitrate and ammonia forms and phosphorus is analyzed as orthophosphate.
- Electron acceptors — groundwater samples should be collected for dissolved oxygen analyses. Depending on project and site-specific circumstances, determination of levels of substitute electron acceptors, nitrate, manganese, iron, and sulfate should also be made.
- By-products of bacterial activity — concentrations of carbon dioxide, a by-product of both aerobic and anaerobic metabolism, and methane gas, which is produced by methanogenic bacteria in anaerobic metabolism should be monitored.
- Total organic carbon (TOC) — knowledge of the TOC content of the aquifer matrix is important in sorption and solute-retardation calculations. The average TOC concentration from the most transmissive zone in the aquifer should be used for retardation calculations.
- Site geology and hydrology — this information should also be available from earlier site assessment and hydrogeological reports.
- Soil conditions — soil and/or groundwater samples should be collected and analyzed for pH, moisture content (soil samples only), temperature, and conductivity.
- Groundwater chemistry — groundwater chemistry is important in determining the success of the final bioremediation system and for predicting O & M problems at the site. Because oxygen is typically introduced as the electron acceptor, precipitation of iron and manganese can be expected in aquifers with notable concentrations of dissolved iron and manganese. The proliferation of iron-oxidizing bacteria can also be a serious problem. As such, concentrations of iron and manganese in their dissolved and insoluble forms should be collected.

  Based on the authors' experience, samples for dissolved iron and manganese measurements should be filtered on-site. This process should be conducted carefully so as not to aerate the samples and increase the insoluble content. Field analyses should also be performed for comparison with laboratory results. Groundwater with high hardness content may also cause precipitation problems.

The locations for collecting the samples are just as important. Sampling points should be located both outside and within the known contamination area.

The outside sampling wells will serve as the background monitoring points or control wells. Where possible, they should be located both upgradient and downgradient of the contamination. Control wells should also be located within the plume for comparison purposes. Control wells within the plume can be located downgradient, cross-gradient, or upgradient of the targeted wells.

The targeted well, i.e., the well intended for demonstrating that bioremediation is taking place, should be located close enough to the biosimulation well such that the added electron acceptors and nutrients can reach it during the duration of the feasibility test. The biostimulation well (or trench) is the location where electron acceptors and nutrients are added to the groundwater.

The pilot program will have to take into account the methods for introducing oxygen (or other electron acceptors) and nutrients into the subsurface. Based on site-specific characteristics, oxygen may be provided through bioventing, sparging, bubbler-diffusion, aerated water injection, or hydrogen peroxide injection. Soluble inorganic fertilizers are used as sources of nitrogen and phosphorus. Micronutrients are usually not provided for a pilot-scale test. Existing wells or a combination of existing and new wells may be used.

Just as in AS and combined SVE/AS tests where helium is used as a tracer to determine flow pathways of sparged air, tracers are also utilized in bioremediation bench-scale tests. Besides tracking airflow pathways and travel times, tracers can also indicate the utilization rate of oxygen. Where air sparging is used to introduce oxygen, helium is injected together with the air supply, much as in SVE/AS tests.

Measurements of dissolved oxygen and helium amounts are taken at the monitoring wells. The rate of oxygen depletion relative to helium depletion is an indication of the microbial activity between the injection well and the monitoring wells. However, a portion of the oxygen may be reacting with other reduced substances, such as iron (II). Therefore, the relative rate of oxygen depletion should be compared with that occurring at a background monitoring well. Where hydrogen peroxide is used as the method of oxygen introduction, bromide is used as the tracer. Bromide is added to the hydrogen peroxide feed tank and the mixture is circulated through the contaminated zone.

6.4.2 Data Evaluation

Data collected before and after the pilot test or before and during the final bioremediation process are analyzed and compared. Several parameters will be used to evaluate the effectiveness of the bioremediation process.

- Contaminant analyses: Contaminant concentrations in soil and/or groundwater samples collected from targeted wells are expected to decrease as bioremediation progresses. A demonstration of a definite decreasing trend in the concentrations of these samples will provide the strongest evidence that bioremediation is feasible.

In the course of a short pilot test, however, significant differences between initial and final concentrations may not be observed. In this case, depending on the results of the other parameters analyzed, it may be necessary to prolong the test. A difference of 20 to 50% in contaminant reduction levels between the targeted well and control well can be used as a rule of thumb (U.S. EPA, 1991b). This means that if the contaminant concentrations in the control well indicated a 10% reduction over the duration of the study, the targeted well should achieve at least a 30 to 60% reduction.

Contaminant concentrations in control wells located within the plume can exhibit decreases, increases, or remained unchanged. Typically, they demonstrate no trends, especially in studies conducted over a short duration. It is important to remember that the objective of the treatability study is to show that bioremediation can treat the contaminants on-site, and not to ascertain whether it can meet cleanup goals.

- Microbial activity: Baseline bacteria plate counts taken just before the bench test are an indication of the biodegradation potential of the site. Standard plate count methods typically report bacteria population densities in colony-forming units per gram of soil (cfu/gm) for soil samples, and in cfu/mL for groundwater samples.

  For petroleum hydrocarbon-impacted sites, microbial analyses generally include a total heterotrophic plate count and a petroleum-degrading bacteria analysis using the most probable number (MPN) method. Increases in microbial counts are considered indicative of stimulated biodegradation. However, it should be noted that bacteria counts seldom demonstrate distinct trends, especially over a short duration. Because of various biological, chemical, and physical conditions, including the fact that bacteria population proliferates exponentially, this criteria should be used discriminately.
- Electron acceptor concentration: In aerobic bioremediation, oxygen concentrations are monitored. The trend to look for in the targeted wells are an increase in the dissolved oxygen concentration. If initial contaminant concentrations are considerably high, a noticeable increase in dissolved oxygen may not be readily apparent.
- By-product concentrations: Increases in carbon dioxide and methane gas concentrations are an indication of bacterial activity in the subsurface. Under normal circumstances, methane level in the soil is approximately 6 mg/L, thus concentrations higher than that can be considered to be indicative of anaerobic bacterial activity and biodegradation.
- Inorganic concentrations: High concentrations of dissolved iron and manganese may be indicative of potential precipitation problems.

6.5 SYSTEM DESIGN

The design of a bioremediation system involves several steps, including:

- Determining what needs to be added to the subsurface to stimulate microbial activity and in what quantities.
- Predicting changes in the chemistry and microbiology in the subsurface as a result of biostimulation and incorporating provisions in the design to counteract them.
- Designing delivery systems.
- Implementing a monitoring program, which serves both as a measure of the success of the program (for regulatory purposes) and as an ongoing information collection system for adjusting the operation of the system, e.g., nutrient requirements — deficient or excessive.

A summary of all findings and conclusions established to-date should first be compiled. These include site geological and hydrogeological data, extent of contamination reports, feasibility study results, and cleanup requirements (which, quite likely, will be the overriding control factor). Design objectives can then be defined based on this information.

If an aggressive approach is desired for remediating contaminated groundwater, the system will probably incorporate wells and trenches to hydraulically control groundwater flow. Above-ground units for treating the extracted groundwater and returning nutrient-amended and pH-adjusted water to the aquifer will be required. Sparge wells will need to be installed for providing oxygen; perhaps along with other oxygen delivery systems, such as hydrogen peroxide and oxygen-releasing "socks."

If a less aggressive approach is preferable, a groundwater recovery system may not be necessary. If containment of the migrating plume is all that is required, then perhaps a "bio"-barrier wall or trench will be sufficient. In certain cases, remedial installations may not be required, and intrinsic bioremediation is acceptable as the treatment alternative. In this case long-term monitoring will be the only remedial measure implemented.

Based on the investigations and pilot-scale treatability tests performed, design parameters can then be identified. For vadose zone systems, these include determining answers to the following questions:

- What ROI can the bioventing wells achieve?
- What pressures and flowrates are we looking at?
- At what rates will we be initially adding nutrients, if at all?

For groundwater systems, design parameters may include:

- What is the ROI of the pumping well?
- What is the recovery rate?
- Nutrient amendment rates?

The final detailed design is essentially a twofold process. The first involves the determination of the quantities of electron acceptors and nutrients required by the microorganisms, followed by the selection and design of the delivery systems. The second involves the determination of the locations of the delivery systems and the design of other treatment technologies to complement the bioremediation system. The system monitoring program will essentially follow the NRC procedures.

6.5.1 Provision of Microorganism Requirements

From the results of the pilot tests, determinations can be made regarding the amount of electron acceptors and nutrients that must be added to the site to stimulate microbial activity. In some cases, it may not be necessary to provide certain nutrients if continuous testing indicates that the particular nutrients are not lacking at the site. For most sites, however, electron acceptors, especially oxygen, need to be provided to the system. Dissolved oxygen levels in hydrocarbon-impacted water-bearing zones can be as low as 0.1 to 0.2 mg/L, with a corresponding high carbon dioxide range of 30 to 150 mg/L.

The objective here is to balance oxygen supply with oxygen demand. Initial design assumptions for calculating oxygen requirements are generally based on the "three pounds of oxygen are needed to convert one pound of hydrocarbon to carbon dioxide and water" ratio. This rule-of-thumb number is derived from mass balance calculations based on the stoichiometry of BTEX biodegradation. In the previous section, the biodegradation of toluene was described by the following equation:

$$C_7H_8 + 9O_2 \rightarrow 7CO_2 + 4H_2O$$

From the above equation, the mole ratio of oxygen to toluene is 9:1, i.e., 9 moles of oxygen are required to serve as electron acceptors by the microbes to metabolize one mole of toluene. The molar masses of toluene and molecular oxygen are

$$C_7H_8 = 7(12) + 8(1) = 92$$

$$O_2 = 2(16) = 32$$

From the mole ratio and molar masses, the mass ratio of oxygen to toluene is then:

$$32(9)/92(1) = 3.13$$

The previous stoichiometric calculations can be repeated for benzene, where the reaction is described by the following equation:

$$C_6H_6 + 7.5O_2 \rightarrow 6CO_2 + 3H_2O$$

The molar mass of benzene is 78, providing a mass ratio of oxygen to benzene of:

$$32(7.5)/78(1) = 3.08$$

Similar stoichiometric calculations can be completed for ethylbenzene and xylene, yielding a mass ratio of 3.17 in both cases. Assuming equal distributions of benzene, toluene, ethylbenzene, and xylene concentrations in the BTEX plume (which is never the case), the average mass ratio of oxygen to BTEX is 3.14:1. Assuming the mass ratio of oxygen to BTEX to be somewhat equivalent to the mass ratio of oxygen to petroleum hydrocarbons, the general guideline of "three pounds of oxygen are needed to degrade one pound of hydrocarbon" is thus used. To determine the amount of oxygen required, the total masses need to be quantified.

The next step is to estimate the amount of petroleum hydrocarbons in the subsurface. One gallon of gasoline weighs approximately 6.2 pounds (specific gravity approximately 74%). Multiplying the pounds of hydrocarbons by a factor of three would provide initial estimates of oxygen requirements to remediate the entire site. However, since oxygen is utilized in almost all biological respiration, this number will be on the low side. Additionally, oxygen will be chemically consumed as it oxidizes the reduced forms of minerals to their insoluble forms.

A very rough approximation of the cleanup time can thus be estimated based on the total amount of oxygen needed to remediate the site and the rate of oxygen supply. "Rough approximation" is emphasized here; there are too many variables and inherent complexities in the overall equation. Perhaps of utmost importance is the efficient utilization of the oxygen by the petroluem degrading bacteria. This is a function of several factors, including the efficient distribution of oxygen within the plume, competition for the oxygen by other types of bacteria, oxidation of inorganics, and the oxidation of other organic carbon materials present in the subsurface. The rate of oxygen supply itself is a function of the site conditions and subsurface characteristics. In bioventing and air sparge systems, i.e., mechanical delivery systems, the allowable injection pressure or vacuum (in some bioventing systems) can be a limiting factor. In these cases, pure oxygen as opposed to air may be used.

For most gasoline-impacted sites, typically only the macronutrients nitrogen and phosphorus are analyzed for. Initial estimates for nutrient addition is usually based on pilot test results. The key is to ensure that there will be sufficient nutrients for the microorganisms, especially during their growth phase. The objective is to maintain a favorable environment for the microbes and not create stressful conditions. Ongoing monitoring during the life of the bioremediation project is the best way of adjusting nutrient requirements.

There are too many complex and interrelated variables and conditions taking place in the subsurface to allow stoichiometric calculations to predetermine exact nutrient requirements. Estimates based on the pilot-scale treatability test area, for all practical purposes, reliably accurate enough to design the system and project operations and maintenance costs.

6.5.2 Design of Delivery Systems

The design of nutrient and oxygen delivery systems can be broadly divided into vadose zone and saturated zone systems. Vadose zone systems may include bioventing wells, blowers, nitrous oxide tanks, infiltration galleries, and nutrient tanks. Components of groundwater systems may include sparge wells, blowers, injection wells, injection pumps, infiltration galleries, recovery wells, submersible pumps, above-ground groundwater treatment units such as air stripping towers and low profile towers, barrier "walls," nutrient tanks, pH tanks, batch mixing tanks, and a programmable logic control (PLC) system.

6.5.3 Design of a Bioventing System

In the bioremediation of vadose zone contamination, the three major control parameters are oxygen supply, moisture maintenance, and nutrient supply. Oxygen supply is generally provided through bioventing of the soils (see Figure 6.6). Oxygen (air) is mechanically introduced into the unsaturated zone via forced pressure (air injection) or vacum application (similar to SVE). There is a very fundamental difference between bioventing and SVE; whereas SVE depends on volatilization as its principal mode of remediating hydrocarbons, the goal of bioventing is to achieve treatment through biodegradation of the contaminants.

To prevent or minimize volatilization and off-site migration of possibly impacted soil vapors, the blower pressure is kept low. Infiltration galleries can be used to provide both nutrients and moisture content. Nitrous oxide can be injected into the subsurface along with air to provide nitrogen. The nitrous oxide tank is connected to the suction side of the blower and its flow regulated as appropriate for the site conditions.

The design of a bioventing system is similar to a typical SVE design. Like SVE, a pilot test should first be completed to confirm its feasibility, and also to establish design or operational parameters, such as injection pressure or vacuum applied, air flowrate, radius of influence (ROI), and the existence of preferential pathways. The major difference is the use of monitoring wells around the perimeter of the site to detect any off-site migration of hydrocarbon-impacted vapors. In addition, bioventing represents a less aggressive approach than SVE.

Cleanup time typically takes longer and extensive monitoring is recommended. However, it does not require above-ground treatment and is generally not as costly to install as SVE. Bioventing is applicable at sites where the soil type provides sufficiently high permeability and homogeneity. In addition, it should work well at sites where the contaminated zone is well defined, both laterally and vertically. At such locations, the ideal design would be a bioventing well(s) located in the center of and screened across the impacted area.

6.5.4 Design of Air Sparge (Biosparging) Wells

The design of air sparge wells to provide oxygen for bioremediation in the saturated zone is similar to that for SVE/AS systems, except that the injection pressure and flowrate is generally kept lower. A typical range for the injection pressure is 3 to 10 psi (depending on local hydrostatic head), whereas a typical flow range is approximately 1 to 3 cubic feet per minute (CFM).

The screened interval or slotted section of the sparge well is generally located below the contaminated region of the saturated zone. The screen interval is typically between 1 to 2 feet in length. As in SVE/AS system designs, a pilot test should first be performed to determine its feasibility at each specific site, and also to establish design and operational parameters, such as injection pressures, flowrates, radius of influence, and the existence of preferential pathways.

6.5.5 Design of Groundwater Reinjection Wells

Injection wells are used in bioremediation for injecting nutrients and/or hydrogen peroxide/other substitute electron acceptors into the saturated zone. In addition, injection wells are also used for injecting aerated water. The design criteria for injection wells are similar to those for pumping wells. Because clogging of injection well screens is a major problem, screen length and screen open areas are critical design considerations and should be optimized. In general, for equal volumes of water, the screen length for an injection well should be twice that of a pumping well (Driscoll, 1986).

6.5.6 Design of Infiltration Galleries

The two primary considerations in designing an infiltration gallery are (1) even distribution of the water to the area of contamination, and (2) achieving equilibrium conditions, i.e., the gallery should be able to store and transmit water to the subsurface at the same rate that nutrient-enriched water is being added to it to prevent flooding.

Infiltration galleries provide certain advantages over injection wells. Because infiltration galleries work on the basic principle of allowing water to percolate through the soil, they can be used to provide nutrients and moisture to the vadose zone. In addition, they have greater coverage and, thus, can provide a more uniform distribution of nutrients and electron acceptors to the water-bearing zone. However, it should be noted that infiltration galleries can, under specific conditions, potentially transport contaminants sorbed in the vadose zone to the groundwater. Unless vadose zone contamination has been removed or a concurrent groundwater treatment system is operational, infiltration galleries should be used with caution.

Infiltration galleries can be used alone or in conjunction with a downgradient pump and treat system. The following examples illustrate the differences in the design for both types of installations.

Example: Assume the following:
A flowrate of 2 gpm is supplied to the gallery; the underlying native material is homogeneous and has a hydraulic conductivity of 25 gpd/ft^2 and a porosity of 0.3; hydraulic gradient of flow in the water-bearing zone is 0.05; the groundwater table is approximately 30 ft below surface grade.

Let A = area of the gallery (ft^2)
Flow in = 2 gpm = 2880 gpd
Flow in per unit area of gallery = 2880/A gpd/ft^2
Flow out is calculated using Darcy's equation,

$$v = Ki$$

where v = specific discharge, which is the quantity of water flowing in unit time through a unit cross-sectional area of soil at right angles to the direction of flow

$$v = 25 \text{ gpd/ft}^2 \times 0.05 = 1.25 \text{ gpd/ft}^2$$

Actual velocity, v_s = 1.25/0.3 = 4.2 gpd/ft^2
From mass balance, flow in = flow out

$$2880/A \text{ gpd/ft}^2 = 4.2 \text{ gpd/ft}^2$$

The minimum cross-sectional area of the infiltration gallery required to prevent flooding,

$$A_{min} = 2880/4.2 = 685 \text{ ft}^2$$

Using a 5-ft-wide trench, the minimum length of the trench should be approximately 137 ft.

Example: In the second situation, the difference in the calculation procedure will be the addition of a steeper hydraulic gradient due to groundwater pumping (Figure 6.10). Assume the flowrate is 20 gpm

Flow in = 10 gpm = 14,400 gpd
The hydraulic gradient, i = Dh/l,
With h = 40' and l = 200'.

$$i = 40/200 = 0.2$$

$$v = 25 \times 0.2 = 5 \text{ gpd/ft}^2$$

$$v_s = 5/0.3 = 16.7 \text{ gpd/ft}^2$$

From mass balance, flow in = flow out,

$$14,400/A = 16.7 \text{ gpd/ft}^2$$

$$A = 862 \text{ ft}^2$$

Assuming a 5-ft-wide trench, a gallery length of 172 ft will be required to prevent flooding. If there is insufficient space to the entire length of the gallery, then the flowrate will have to be reduced. In the case of a pump and treat system, a portion of the treated water may have to be routed elsewhere, such as to the sewer.

A typical infiltration gallery cross-section is shown in Figure 6.11. Clean-outs should be provided to provide access for removal of clogged material.

6.5.7 Design of Containment Systems

The primary objective of a containment system is to prevent dissolved contaminants in the groundwater from migrating off-site or from reaching sensitive downgradient and, in some cases, cross-gradient receptacles. Typical containment systems include a trench design or a series of vertical wells located downgradient of the contamination source.

A trench system will consist of a trench with vertical wells and horizontal nutrient galleries installed within the trench and backfilled wih highly permeable material. Because the material within the trench is much more permeable than the surrounding aquifer, movement of oxygen and nutrients within the trench will occur more readily and effectively.

Besides permeability, grain size distribution may also play a significant role in the prevention of biofouling. Studies have shown that high-porosity materials that are well-graded, i.e., with widely distributed grain sizes, especially with small diameter pore size ranges, are much more susceptible to biofouling than high porosity material with a limited pore size range. Thus, a free-draining material may likely be more amenable to bioremediation than a well-graded material.

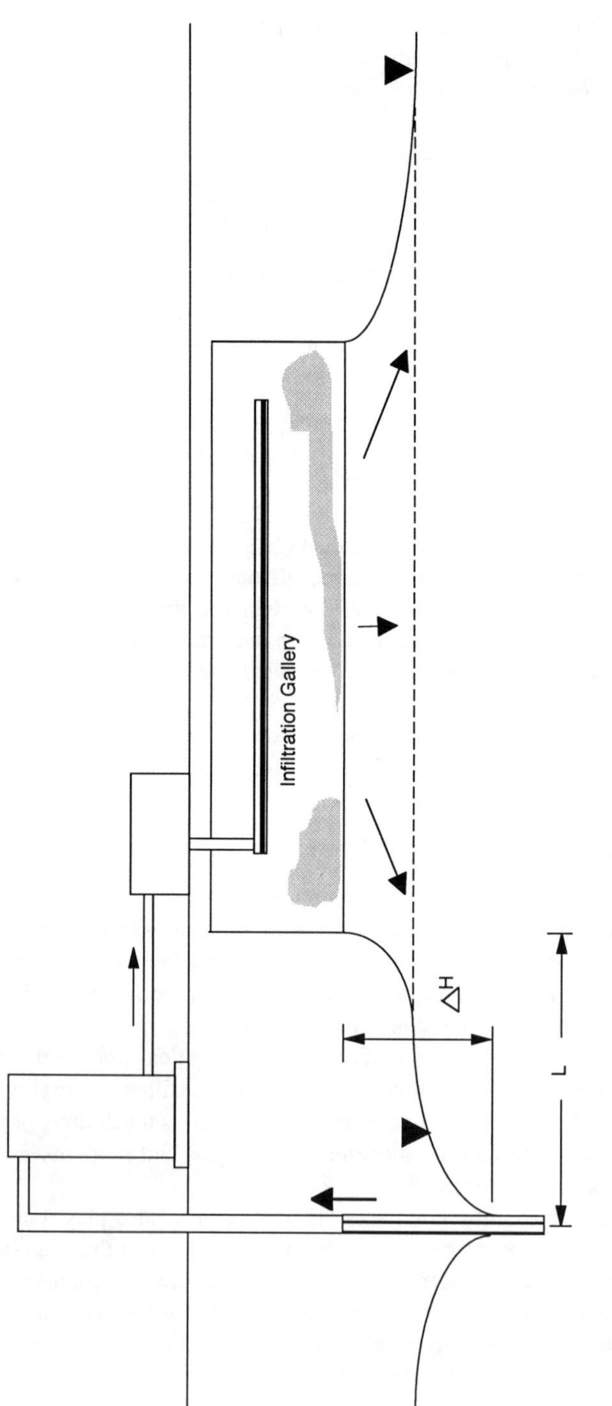

Figure 6.10. Groundwater recirculating system with recovery well and infiltration gallery.

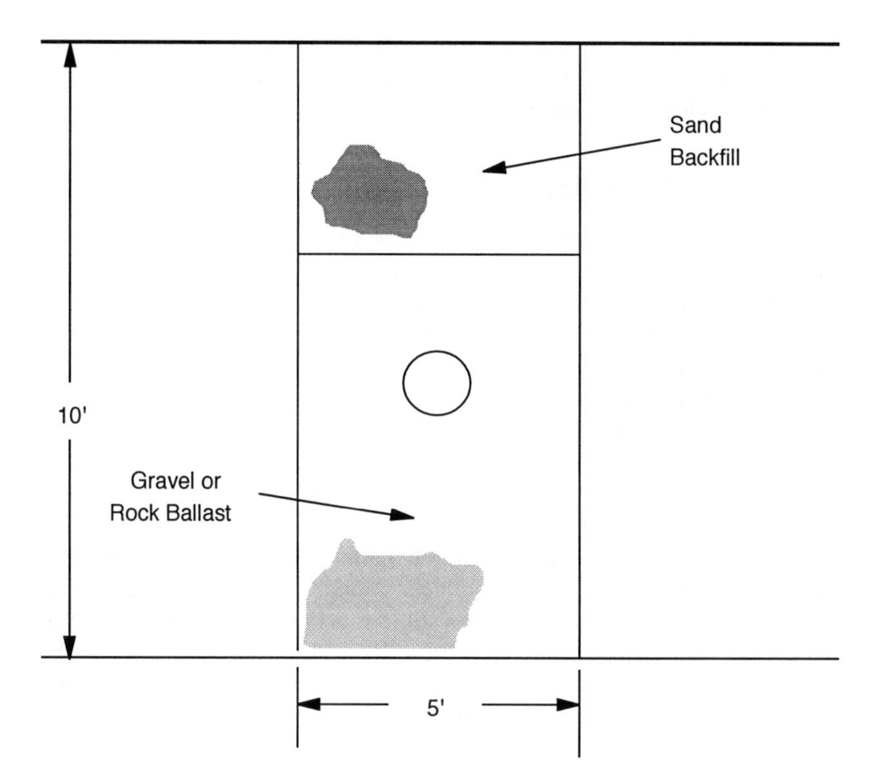

Figure 6.11. Cross-section of infiltration gallery.

Another alternative for designing biological containment systems is to install a series of vertical wells constructed somewhat perpendicular to the groundwater flow direction. This configuration has advantages and disadvantages compared to the trench design. Obviously, it has less oxygen and nutrient transportation capabilities than a barrier trench. However, it can be installed deeper much more economically than digging a vertical trench.

Three issues that typically arise in containment design are

- Mechanism of oxygen delivery
- Depth of trench or wells
- Center-to-center spacing of vertical wells

Inorganic compounds that release oxygen slowly when in contact with water are especially useful in containment systems. Barrier systems do not require a high amount of oxygen; the oxygen also does not need to be transported to great distances. As such, patented products such as Regenesis' Oxygen Releasing Compound (ORC) socks can be installed in the vertical wells to form an oxygen barrier.

For LNAPL such as gasoline, the depth of the trench or wells is not as critical. The spacing of the wells can be determined through a pilot test or through modeling of the aquifer parameters.

6.5.8 Above-Ground Units

Air compressors, blowers, and air pressure and flowrate controllers are similar to those used in SVE/AS systems and have been described in previous sections. Treatment systems for treating extracted groundwater will be described in Chapter 7. The treated water is reinjected into the aquifer after nutrients and/or hydrogen peroxide has been added. Nutrient-amended injection water systems typically consist of a batch tank that incorporates a self-metering pH adjustment tank and a self-metering nutrient tank.

Provisions should also be made for the treated groundwater to be discharged elsewhere, such as to a stream (which would require a National Pollutant Discharge Elimination System permit) or to a sanitary sewer. A three-way valve system can be used for controlling the discharge of treated groundwater to the reinjection system or local publicly owned treatment works (POTW)/stream. Groundwater systems with extraction, reinjection, and/or biosparge wells should be monitored and run by a PLC. The system should be designed with intrinsic safety features such as automatic shut down in the event that the groundwater treatment unit malfunctions.

6.6 INTRINSIC BIOREMEDIATION

Intrinsic bioremediation or natural attenuation is gaining acceptance. Not long ago the concept and approach of allowing nature to heal herself was considered too unreliable and not pro-active enough in the management of contaminated properties. However, with the push for risk-based corrective action (RBCA) and "sensible" remediation, and probably, more than any other combination of reasons, the economics of "all-out" cleanup goals finally catching up with available cleanup funds, natural attenuation is finally considered a legitimate site remedial alternative.

In what is perhaps the strongest endorsement for intrinsic remediation to-date, the Lawrence Livermore National Laboratory, in a report to the California State Water Resources Control Board (SWRCB), recommended utilizing intrinsic or passive remediation as a remedial alternative whenever possible. The same report also criticized present petroleum hydrocarbon cleanup goals, which place a ridiculously high premium on the value of impacted groundwater, and recommended that RBCA be implemented into the cleanup decision-making process.

While natural attenuation processes include biodegradation, dispersion (plume dilution), advection, sorption and volatilization, it is now generally recognized that the biodegradation of petroleum hydrocarbons by indigenous microorganisms is the primary mechanism whereby intrinsic remediation is achieved.

It is also the only mechanism that actually destroys the contaminants, as opposed to transporting them elsewhere or converting them from one form to another.

It is important to remember that intrinsic remediation is not equivalent to a "no action" course of response. While it is true that no active remediation system will be installed, intrinsic remediation requires extensive data collection, numerical model simulation, and long-term monitoring to support its implementation. A strict field protocol must be developed for the specific site and adhered to. The Air Force Center for Environmental Excellence and several other organizations have been primarily responsible for the development of intrinsic bioremediation guidelines and promoting its applicability nationwide.

6.6.1 Concept of Intrinsic Bioremediation

The key to an intrinsic bioremediation program is gathering data to provide evidence that biotransformation is taking place and is capable of remediating the site. There are essentially three ways to provide evidence to support intrinsic bioremediation: demonstrate decreasing contaminant concentrations over time; documenting loss of electron acceptors, both aerobic and anaerobic, over time; and illustrating an increase in levels of by-products associated with the metabolic processes of petroleum-degrading microbes. These indicators of natural attenuation are explained in further detail in the following sections.

6.6.1.1 Document Loss of Contaminants

The documentation of reduction in contaminant levels as a function of time is routine in the monitoring of remediation projects. As explained earlier, this task is made more difficult in bioremediation systems because contaminant losses in groundwater samples can be attributed to other mechanisms besides biotransformation. These mechanisms include volatilization (transfer of contaminants from the dissolved phase to the vapor phase), sorption (reduction in groundwater concentrations as contaminants sorp to the soil matrix), advection (transportation of contaminants by groundwater, generally in a downgradient direction), and dispersion or diffusion (dilution of the contaminant plume as the chemical compounds are mixed with groundwater along the edges of the plume and also through recharge of the aquifer).

While it is generally recognized that natural biodegradation is principally responsible for decreasing levels of contamination, it is still a necessary component of bioremediation projects to provide evidence that the indigenous microorganisms are responsible for this loss. This task is compounded further in intrinsic bioremediation projects. The reason for this is simple; in engineered bioremediation systems, the addition of electron acceptors and nutrients creates a highly favorable environment for expediting the biotransformation process. This creates a highly apparent decrease in contaminant levels in the areas where electron acceptors and nutrients were added, easily demonstrating bioremediation to be responsible. In intrinsic bioremediation, it may not be as easy to differentiate between biotransformation of the contaminant from the mechanisms of dispersion, sorption, volatilization, and advection.

Several techniques have been developed to determine if contaminant loss is due to intrinsic biodegradation.

- Use of a conservative tracer or recalcitrant chemical — A conservative tracer is a component of the gasoline make-up that is not susceptible to bioremediation, i.e., it is not biodegradable. Thus, any decrease in its concentration between two points, (e.g., point A and a downgradient point B), is most likely due to nonbiodegradation mechanisms, i.e., volatilization, sorption, advection, and dispersion. Certain components of petroleum products may be recalcitrant. Two chemicals have been suggested in the literature, methyl tert butyl ether (MTBE) and trimethylbenzene (TMB), for use as conservative tracers. 1,2,4- and 1,3,5-TMB concentrations may be determined by SW-846 Methods 8240 and 8260.

  Because the conservative tracer compound and BTEX both have similar chemical properties, their loss due to nonbiodegradation mechanisms should be proportional. Thus, by determining the BTEX loss due to nonbiological means and subtracting that from the total BTEX loss, the portion attributed to intrinsic biodegradation can be calculated. It should be noted that TMB is not 100% recalcitrant and *is* biodegradable, except at a slower rate than BTEX. The degree of recalcitrance of TMB is site-specific and its use as a conservative tracer in support of intrinsic bioremediation must be considered on a site-by-site basis.

  To determine the percent loss due to nonbioremediation mechanisms, the following equation can be used:

$$\text{Percent Loss (non} - \text{bio)} = \frac{CT_A - CT_B}{CT_A} \tag{6.1}$$

  where CT_A = conservative tracer concentration at point A
  CT_B = conservative tracer concentration at downgradient point B

  Loss due to nonbioremediation mechanisms in the targeted contaminant can be determined by the following equation:

$$\text{Loss due to Non} - \text{Bio in Targeted Contaminant} = \frac{CT_A - CT_B}{CT_A}\left(TC_A\right) \tag{6.2}$$

  where TC_A = targeted contaminant concentration in point A

Loss due to intrinsic bioremediation in targeted contaminant is given by

$$\left(TC_A - TC_B\right) - \frac{CT_A - CT_B}{CT_A}\left(TC_A\right) \tag{6.3}$$

where TC_B = targeted contaminant concentration in downgradient point B

Percent loss due to intrinsic bioremediation in targeted contaminant,

$$= \frac{\text{Loss due to intrinsic bioremediation in targeted contaminant}}{\text{Total loss in targeted contaminant}} \tag{6.4}$$

Samples should be collected from a minimum of three points along a straight line parallel to the direction of groundwater flow. Depending on site conditions, the ideal location for the line will be one running through the source of the contamination. Point A will ideally be located close to the area of highest contamination, and points B and C at downgradient locations of decreasing contaminant concentrations.

- Demonstrate mass loss over time — This is a quantitative method that requires extensive sampling. The objective is to collect samples from enough monitoring wells to contruct accurate TPHg and BTEX contour maps. These contour maps should have well-defined nondetect lines. These maps should be plotted on a regular basis, say, every quarter, over a period of at least one year.

  Based on the concentrations of the contours and the calculated volumes of groundwater within the contaminant plume, it can be shown that the total mass of TPHg and BTEX in the plume is decreasing. Because it is a quantitative method, it does not matter if the plume is expanding, i.e., increasing in length and size over time. The determining factor is a demonstration of decreasing total TPHg and BTEX mass over time. Because the mass calculations encompass the entire plume, any mass reductions cannot be attributed to advection or dispersion.

  There are several inherent problems with this method. If the mass loss is minor, it may be attributed to volatilization or sorption (if the plume is still expanding). Additionally, if the vadose zone soil around the contaminant source area is still impacted, contaminants will be continuously introduced into the groundwater, thereby masking any mass decreases due to intrinsic remediation. By the same token, contaminants sorped in the smear zone will also be a continuous source of contaminant "recharge." Obviously, for this method to be effective in demonstrating intrinsic bioremediation, these factors must be accounted for.

6.6.1.2 Document Loss of Electron Acceptors

Measuring dissolved oxygen concentrations and those of other electron acceptors can provide a key indicator of natural attenuation. Reduced oxygen, nitrate, and sulfate concentrations within the plume relative to their background concentrations are considered to be strong evidence of intrinsic bioremediation. Aerobic biodegradation occurs primarily around the edges of a petroleum hydrocarbon plume because of the availability of "fresh" supplies of oxygen in this region of the plume.

In general, aerobic biodegradation is readily taking place in the plume areas where oxygen concentrations are greater than 2 mg/L. In areas with oxygen concentrations below 0.1 mg/L, anaerobic processes will predominate. Between 0.1 and 2 mg/L oxygen levels, sometimes called hypoxic conditions, aerobic biodegradation will take place but will be severely limited by available oxygen. In addition, at the lower end of the hypoxic category, there is some indication that the biodegradation of BTEX under nitrate-reducing conditions may be taking place (Mikesell et al., 1993).

- Dissolved oxygen — Areas with elevated TPHg and total BTEX concentrations should have depleted dissolved oxygen (DO) concentrations. This typically occurs in the region near the center of the plume. These DO concentrations should be compared with other DO levels from within the plume and also from background monitoring wells. Higher DO levels in the background monitoring wells will be a strong indication of intrinsic bioremediation. However, because of the low DO concentrations in the center of the plume, aerobic biodegradation may no longer be active in that region of the plume.

  Similar low DO levels in the background monitoring wells may suggest that the low DO readings in the center of the plume are not an indicator of aerobic bioremediation at the site. However, this is not conclusive. Low DO levels in the background may be due to other factors. Anaerobic aquifers, i.e., where reducing conditions predominate, are not uncommon.
- Nitrate — Where intrinsic bioremediation is occurring, areas with elevated TPHg and total BTEX concentrations will likely have depleted nitrate concentrations.
- Sulfate — Through the microbially mediated process of sulfanogenesis, sulfate is used as an electron acceptor. Areas with elevated TPHg and total BTEX concentrations will likely have depleted sulfate concentrations.

6.6.1.3 Document Increasing By-Products of Metabolism

The production of carbon dioxide, hydrogen sulfide, and methane and the accumulation of dissolved iron are additional indicators of biological attenuation at intrinsic bioremediation sites.

- Carbon dioxide — Metabolic processes occurring during the biodegradation of petroleum hydrocarbons leads to the production of carbon dioxide. Aquifers under reducing conditions typically have low levels of oxygen and high concentrations of carbon dioxide. However, accurate determination of carbon dioxide produced as a result of biodegradation is difficult because carbonate in groundwater serves as both a source and sink for free carbon dioxide. If the carbon dioxide produced during metabolism is not removed by the natural carbonate buffering system of the aquifer, carbon dioxide concentrations higher than background may be observed.
- Methane — The presence of methane in groundwater is indicative of strong reducing conditions. Because methane is not present in petroleum fuels, the presence of methane in groundwater above background concentrations is indicative of microbial degradation of fuel hydrocarbons.
- Ferrous iron — Ferrous or dissolved iron is formed when ferric or insoluble iron is used as an electron acceptor in the anaerobic biodegradation of petroleum hydrocarbons. The trend to look for is areas with elevated TPHg and total BTEX coinciding with elevated ferrous iron concentrations. This condition by itself will not be an indicator of intrinsic bioremediation. This is because elevated levels of ferrous iron are fairly common, especially under reducing conditions in the aquifer.

6.6.2 Will Intrinsic Bioremediation Be Sufficient?

While the techniques described in the previous section may provide sufficient evidence of ongoing intrinsic bioremediation, it is still necessary to use the data to demonstrate that natural attenuation is capable of remediating the site or at least limiting further migration of the plume. This is typically demonstrated by performing mass balance calculations on equations describing the stoichiometry of BTEX degradation using electron acceptors under both aerobic and anaerobic conditions. The objective is to show that the total assimilative capacity of the aquifer is capable of biodegrading the total mass of BTEX present in the plume. This mass balance approach uses an analysis of BTEX plume characteristics along with groundwater data and plume configurations.

From background electron acceptor levels, calculations can be made to determine levels of BTEX that can be biologically removed.

- Dissolved oxygen — From a previous section, it was stoichiometrically shown that the average mass ratio of oxygen consumed to total BTEX degraded is 3.14:1. This means that approximately 0.32 mg of BTEX are biodegraded to carbon dioxide and water for every 1.0 mg of dissolved oxygen consumed. By multiplying the average background dissolved oxygen concentration (in mg/L) by 0.32, the concentration of total BTEX that the aquifer can assimilate aerobically can be semiquantitatively determined.
- Nitrate — The stoichiometric equation for nitrate reduction in the biodegradation of benzene is presented by the following equation,

$$C_6H_6 + 5NO_3^- \rightarrow 6CO_2 + 3H_2O + 2.5N_2$$

From the above equation, the mole ratio of nitrate to benzene is 5:1, i.e., 5 moles of nitrate are required to serve as electron acceptors to biodegrade 1 mole of benzene. The molar masses of benzene and nitrate are

$$C_6H_6 = 6(12) + 6(1) = 78$$

$$NO_3^- = 1(14) + 3(16) = 62$$

From the mole ratio and molar masses, the mass ratio of nitrate to benzene is

$$62(5)/78(1) = 3.97$$

Thus, 3.97 mg of nitrate are required to metabolize one mg of benzene. Similar calculations can be completed for toluene (4.04 mg nitrate to 1 mg toluene), ethylbenzene (4.09 mg nitrate to 1 mg ethylbenzene), and xylene (4.09 mg nitrate to one mg xylene), based on the following stoichiometric equations:

$$\text{Toluene:} \quad 6NO_3^- + C_7H_8 \rightarrow 7CO_2 + 4H_2O + 3N_2$$

$$\text{Ethylbenzene:} \quad 7NO_3^- + C_8H_{10} \rightarrow 8CO_2 + 5H_2O + 3.5N_2$$

$$\text{Xylene:} \quad 7NO_3^- + C_8H_{10} \rightarrow 8CO_2 + 5H_2O + 3.5N_2$$

The average mass ratio of nitrate consumed to total BTEX degraded is thus 4.05:1. This means that approximately 0.25 mg of BTEX are biodegraded for every 1 mg of nitrate consumed. By multiplying the average background nitrate concentration (in mg/L) by 0.25, the concentration of total BTEX that the aquifer can assimilate through microbially mediated nitrate reduction can be semiquantitatively determined.

- Sulfate — By performing stoichiometric calculations as conducted for dissolved oxygen and nitrate, it was determined that the average mass ratio of sulfate to total BTEX is 4.7:1. This is equivalent to the biodegradation of approximately 0.21 mg of BTEX for every 1 mg of sulfate utilized as electron acceptors. By multiplying the average background sulfate concentration (in mg/L) by 0.21, the concentration of total BTEX that the aquifer can assimilate through sulfanogenesis can be semiquantitatively determined.
- By monitoring concentrations of by-products of metabolism, ferrous iron (from iron reduction) and methane (from methanogenesis), similar stoichiometric and mass calculations can be performed to determine the assimilative capacity of the aquifer. It can be shown that 21.8 mg of dissolved iron is produced for every 1 mg of BTEX biodegraded. Thus, the concentration of total BTEX that can be assimilated by the aquifer through iron reduction is obtained by dividing the dissolved iron concentration in the plume by 21.8. Similarly, it can be demonstrated that 0.78 mg/L of methane is produced during biodegradation (methanogenesis) of 1 mg of total BTEX. The concentration of total BTEX that can be assimilated by the aquifer through methanogenesis is calculated by dividing the methane concentration in the plume by 0.78.

The combined total BTEX assimilative capacity of the aquifer is determined by adding the individual concentrations calculated for each electron acceptor. By multiplying this value with the available volume of "fresh" groundwater in the aquifer, a determination can be made on the quantitative assimilative capacity of the aquifer. By comparing this quantity with the calculated mass of total BTEX in the plume, a determination can be made if intrinsic bioremediation alone (without any addition of electron acceptors) is sufficient to remediate the site. Several factors also come into play in making this determination. These include plume characteristics, aquifer use, and the presence and proximity of critical downgradient receptors.

Based on plume characteristics over a period of time, dissolved hydrocarbon plumes can generally be grouped under four broad categories. These four plume configurations are the expanding plume, the steady-state plume, the shrinking plume, and the fluctuating plume.

- The expanding plume — An expanding plume occurs from the time a spill reaches groundwater and continues until equilibrium is reached between source contribution and the mechanisms of intrinsic biodegradation, sorption, and volatilization.

- The steady-state plume — Most contaminant plumes develop until they reach a steady-state condition. Depending on a multitude of factors, this may take a relatively short period of time or a very long time. Critical among these factors are the size of the spill, removal of the source, concentrations of electron acceptors, and hydrogeologic conditions.
- The shrinking plume — A shrinking plume occurs after steady-state conditions have been achieved and provides strong evidence of ongoing intrinsic bioremediation.
- The fluctuating plume — A fluctuating plume is usually the result of periodic changes in the quantities of contaminants entering the aquifer, either at the source or the smear zone. This could be due to unseasonal fluctuations of the water table or changes in the groundwater flow direction.

6.6.3 Implementing Intrinsic Bioremediation: Protocols and Procedures

The design of an intrinsic bioremediation system, unlike a conventional treatment system, resembles more of a scientific study. It involves extensive data collection, correlation of data between different zones of contamination and control points, and comparison of field results with simulated models. The design engineer's most critical role is probably selecting the most appropriate locations for collecting samples and balancing the number of sample locations with a "real-world" budget that is generally not too sympathetic toward costs associated with scientific data and research.

Before proceeding with an intrinsic remediation project, a detailed workplan will first have to be developed. This workplan will need to be clear on its objectives and how it will prove that natural attenuation is occurring. In general, the various stages of an intrinsic remediation project can be summarized as follows:

- Preliminary assessment of intrinsic remediation
- Data collection for additional site characterization
- Intrinsic bioremediation modeling
- Exposure assessment
- Long-term monitoring

6.6.3.1 Preliminary Assessment

Based on existing site characterization data, a preliminary assessment should be made to determine the viability of intrinsic bioremediation. At a minimum, contaminant contours should be available, along with geological and hydrogeological data. In addition, at least one set of DO, nitrate, dissolved iron, pH, and bacteria plate count data should be available. The purpose of the preliminary assessment is to make optimum use of existing site data to establish an initial opinion of the feasibility of intrinsic bioremediation before committing additional funds and proceeding with extensive data collection.

6.6.3.2 Additional Site Characterization

If the preliminary assessment indicates that intrinsic bioremediation is feasible, the next step consists of performing additional site characterization to provide definitive evidence in support of intrinsic remediation. There are two objectives here: (1) to determine if natural mechanisms of contaminant attenuation are occurring at rates sufficient to protect human health and the environment, and (2) to provide sufficient site-specific data to allow prediction of the future extent and concentration of a contaminant plume through numerical groundwater modeling.

A typical scope of work required to complete the additional scope of work is as follows:

1. Data Collection
   a. Soil Parameters
      Soil samples should be collected from the saturated zone and analyzed for the following parameters:
      - Total organic carbon
      - Clay mineral content
   b. Groundwater Parameters
      The following groundwater parameters should be analyzed:
      - Dissolved oxygen
      - Dissolved carbon dioxide
      - Redox potential
      - pH, temperature, and conductivity
      - Alkalinity
      - Nitrate
      - Sulfate and sulfide
      - Ferrous iron
      - Methane
      - TPHg/BTEX
   c. Soil Vapor Parameters
      The following parameters should be analyzed for soil vapor samples:
      - Oxygen
      - Carbon dioxide
      - Methane
      - Hydrogen sulfide

Sample locations should enable a reasonably accurate contour map to be drawn. The frequency of sampling should be adequate to provide comparison. In general, real-world budgets would most likely not allow for the sampling frequencies commonly associated with well-funded research projects.

2. Site-Specific Hydrogeology
   - Hydraulic conductivity
   - Hydraulic gradient
   - Flow direction
3. Definition of potential exposure pathways and receptors
   - Potential migration pathways
   - Downgradient receptors

6.6.4 Intrinsic Bioremediation Modeling

Predicting the extent of plume migration involves estimating two critical values: the rate of contaminant migration, and the rate of contaminant biodegradation. The rate of contaminant migration can be estimated from aquifer characteristics. Biodegradation rates are more difficult to estimate. Many biodegradation models have been developed in recent years, including BIOPLUME, BIOPLUME II, BIOPLUS, and ULTRA. Most of these models utilize some kinetic expression for biodegradation and are generally similar in that they simulate the transport and biodegradation of a number of components in the groundwater. The models differ in the mathematical biodegradation expressions that are used and in the numerical procedures used in solving the equations.

Models have been used to help estimate future migration and attenuation of contaminant plumes at contaminated sites. The models also aid in selecting placement of monitoring wells and sampling strategies. In addition, the models will be able to predict when a steady-state plume will develop.

6.6.5 Exposure Assessment

An exposure assessment identifies the human and ecological receptors that could come into contact with the contaminants and the pathways through which these receptors might be exposed. If a complete exposure pathway is present, and biodegradation rates estimated will likely not prevent exposure, then instrinsic bioremediation alone may not be a viable alternative.

6.6.6 Long-Term Monitoring

Long-term monitoring well locations are selected based on the results of the previous stages of investigation. Likewise, based on calculated advective groundwater velocities for the site, and observed and predicted contaminant migration rates, sampling frequencies can be recommended.

6.7 DESIGN EXAMPLES

6.7.1 Example 1

The following example takes the design engineer from the initial assessment stage, through the review of remedial alternatives, and the pilot-scale test, and finally the conceptual design. It is assumed that site characterization data have been collected and are available.

6.7.2 Scenario

The site is a trucking facility with ongoing operations. Two underground storage tanks were removed and petroleum-contaminated soils and groundwater were evident. One of the tanks was used to store gasoline, the other, diesel. A detailed hydrogeological investigation was conducted and permanent monitoring wells were installed. Soil and groundwater samples collected from on-site borings on the subject property confirmed the presence of petroleum hydrocarbons in the subsurface.

6.7.3 Previous Findings and Conclusions

Geological, hydrogeological, and sampling data compiled to-date provided the following site characterization information:

- Site geology is characterized by a homogeneous 8-ft clay layer above generally sandy material. The sand layer extends to the explored depth of 25 ft below surface grade (bsg).
- Water table observed during drilling operations was at approximately 8 to 8.5 ft bsg. It is unclear if the water-bearing zone is a confined aquifer.
- Soils within the clay layer are relatively "clean," except for those close to the water table around the former underground storage tank complex. The upper 6 to 7 ft bsg had nondetectable levels of TPHg and BTEX. The bottom 1 to 2 ft of the clay layer was impacted at some locations, presumably due to smearing from variations in groundwater elevations.
- Groundwater is impacted with TPHg and BTEX, with the highest concentrations around the tank complex and diminishing radially from it (Figure 6.12).
- Total petroleum hydrocarbons as diesel (TPHd) were not detected.
- A bioremediation feasibility study concluded that bioremediation is feasible at this site. Natural microorganisms are present in sufficient numbers in the subsurface to biodegrade the targeted contaminants.
- By providing oxygen and the macronutrients, nitrogen and phosphorus, the study demonstrated a significant reduction in petroleum

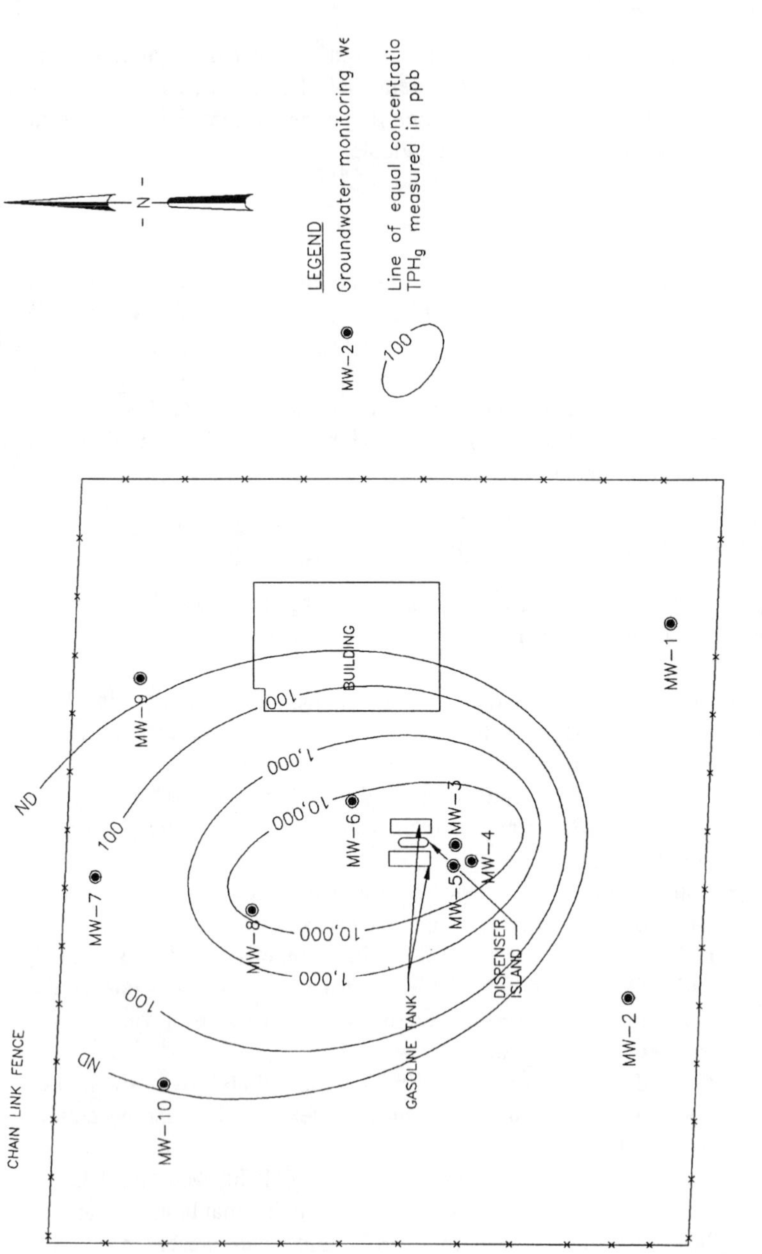

Figure 6.12. TPHg isoconcentrations detected in groundwater.

hydrocarbon concentrations. The lack of nutrients, especially nitrogen, is evident at this site. A successful in situ bioremediation system will require the addition of nitrogen but may not require phosphate addition. Oxygen was provided by oxygen release compounds (ORC) in the feasibility study.

6.7.4 Design Objectives

The proposed remediation program has the following objectives:

1. Reduce concentrations of TPHg and benzene in groundwater moving off-site to acceptable levels.
2. Because vadose zone contamination is minimal, soils treatment will not be required.

6.7.5 Conceptual Design

The proposed containment system consists of a 200-ft-long by 4-ft-wide and 18-ft-deep bioremediation trench laid across the flowpath of the groundwater (Figure 6.13). The trench is located along the north boundary of the site (downgradient from the source) and is designed to form a hydrocarbon-degrading barrier. The trench material, with a higher permeability than the surrounding native materials, is designed to transmit and distribute oxygen and nutrients principally within the trench, although a portion will eventually be carried downgradient.

Oxygen for the microorganisms is provided by ORC. The ORC socks will be installed in vertical wells placed in the trench. The wells are to be 15 feet apart, center-to-center (Figure 6.14 and 6.15). Nutrients (essentially nitrogen and phosphorus) will be delivered via a horizontal pipe laid in the trench. The delivery system for nutrients is simple and nonmechanical. Nitrogen and phosphorus salts will be mixed with water and poured down the gallery pipes at manholes A and B. The same system can also be used to deliver oxygen in the form of hydrogen peroxide pellets. Likewise, the row of vertical wells can also be used as delivery systems for nutrients and hydrogen peroxide.

Monitoring wells are installed both upgradient and downgradient of the containment trench to monitor the efficiency of the trench. In addition, data from these monitoring wells will be used to adjust oxygen and nutrient supplies to the system.

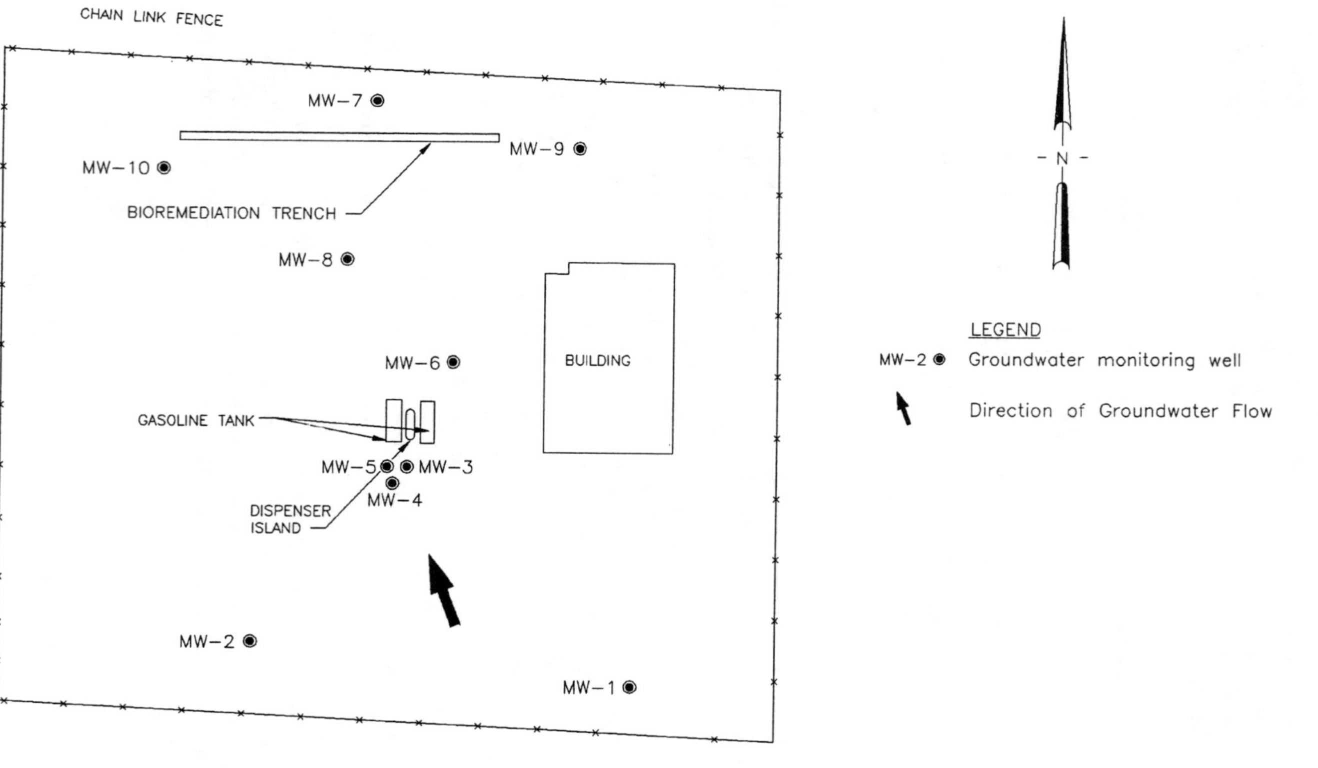

Figure 6.13. Conceptual design for containment system.

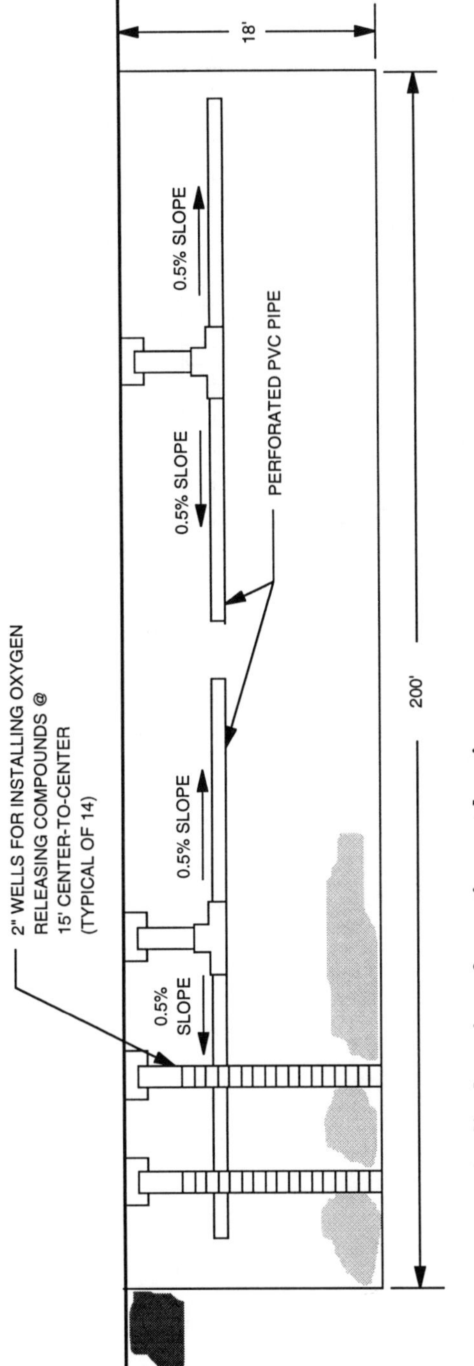

Figure 6.14. Longitudinal section of containment barrier.

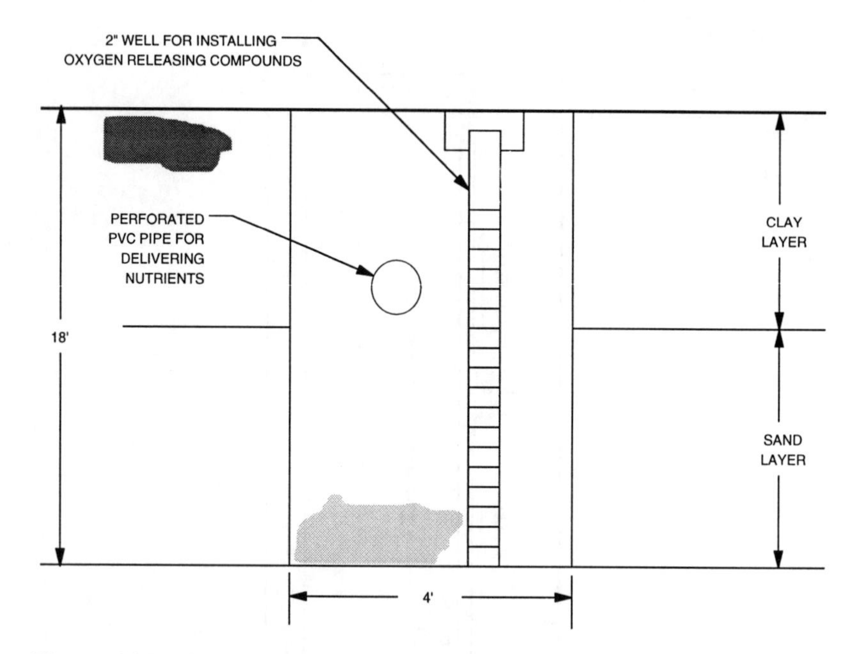

Figure 6.15. Cross-section of containment barrier.

7 DESIGN OF PUMP AND TREAT SYSTEMS

7.1 INTRODUCTION

Up until recently, pump and treat was probably the most commonly deployed groundwater treatment method. However, due to some negative commentaries, pump and treat is no longer widely regarded as the groundwater treatment system of choice. These commentaries mostly pertain to two critical issues in remediation engineering: (1) time to achieve cleanup goals, and (2) effectiveness of the system (volume of water required to be pumped to remove a certain amount of contamination). Additionally, the introduction of air sparging and the advent of in situ bioremediation has provided engineers with a wider array of remedial alternatives.

The basic concept of pump and treat is simple. It utilizes one or more groundwater pumping wells to draw the contaminated groundwater to the surface. The recovered groundwater is in turn then routed to an above-ground treatment unit where the dissolved contaminants are removed. The treated groundwater is then discharged to a water body or a publicly owned treatment works (POTW), or reinjected back into the subsurface. Where the recovered groundwater contains non-aqueous phase liquid (NAPL) or free product, phase-separation is usually conducted prior to treatment. At sites where there are substantial amounts of light non-aqueous phase liquid (LNAPL) to be recovered, and subsurface conditions are suitable, a second pump, usually called the product or recovery pump, is installed above the deeper or depression pump. Above-ground technologies for treating the contaminated groundwater include vapor stripping (air stripping towers and tray aeration units), adsorption (carbon-based or polymer-based vessels), membranes (reverse osmosis), and microorganisms (bioreactors). Techniques that employ vapor stripping will need to treat the off-gas generated. Vapor abatement techniques include oxidizers and carbon adsorption units.

The primary goals of pump and treat systems can be described as follows:

- Prevent or contain contaminant migration to protect the subsurface from further contamination, and to protect downgradient water bodies or structures.

- Restore contaminated water-bearing zones by reducing dissolved contaminant concentrations to acceptable levels.
- Recover LNAPL by drawing free-floating product toward the recovery wells.

Additionally, pump and treat can be used in conjunction with other treatment technologies to enhance remediation results. Two common examples are (1) the use of pump and treat with soil vapor extraction (SVE) simultaneously in dual-phase wells, and (2) the deployment of recovery wells in tandem with injection wells and infiltration galleries in in situ bioremediation systems. In the first case, the drawdown and cone of depression created by the recovery well(s) presents an unsaturated zone below the static water table for vapor extraction by the SVE system. In the second example, the actions of the recovery and injection wells create a circulation of nutrient- and electron acceptor-enhanced groundwater.

Despite some criticisms of pump and treat in the literature, it is still a useful treatment method, especially where there is a significant amount of free product to be recovered. In addition, by creating a hydraulic capture zone, a properly designed pump and treat system can be extremely effective in preventing further migrations of contaminants in groundwater.

7.2 SITE CHARACTERIZATION AND DATA REQUIREMENTS

Data collection and site characterization is a prerequisite for all remedial designs. For pump and treat systems, this is even more crucial than for an SVE/air sparging (AS) system. This is because the success of the system is a function of how accurately the designers and their computer models predict the aquifer will behave when the system is in operation. The information that is entered into the computer is provided by site characterization data. Additionally, unlike SVE systems which are basically a vadose zone remedial technique, pump and treat systems may have significant impact on the local aquifer. Pump and treat systems are also used in decidedly more complicated situations where SVE/AS or bioremediation alone may not be appropriate, such as where there is a thick layer of free-floating product or pools of dense non-aqueous phase liquids (DNAPL) or where plume migration control is required. In order to design an effective pump and treat system for a specific site, the following data requirements must be addressed:

- Aquifer characterization
- Geological characterization
- Extent of contamination
- Contaminant characterization
- Groundwater analyses
- Other site-specific characteristics

Because the contamination plume is constantly moving in the subsurface, the pump and treat system should be installed as soon as possible after site characterization is completed. This is especially true for sites where the subsurface consists of soils with high permeability. In such cases, such as at a recent spill in coarse sand, an interim system may first be installed. The interim pump and treat system serves two purposes: (1) to recover as much product as possible in the shortest time, and (2) to prevent the product from migrating away from the source area. By installing a recovery system early, the amount of product dissolving into the local water-bearing zone and sorping into the soil matrix can be minimized.

7.2.1 Aquifer Characterization

One of the key factors influencing pump and treat system design is aquifer characterization. Aquifer characteristics that play an important role in the design of pump and treat systems include physical parameters such as hydraulic conductivity, aquifer thickness, and storage coefficient; system dynamics, e.g., pumping and recharging rates; and other system characteristics such as groundwater flow directions and hydraulic gradients.

7.2.1.1 Hydraulic Conductivity and Transmissivity

Hydraulic conductivity is a measure of the capacity of a soil media to transmit flow of a specific fluid while transmissivity is the capacity of an aquifer to transmit water. In pump and treat systems, hydraulic conductivity and transmissivity values are used to determine the locations and pumping rates of groundwater extraction wells and estimate the migration rates of contaminant. This information is also useful in calculating the number of wells required and their effective spacing or distance apart.

7.2.1.2 Aquifer Thickness and Plume Depth

At typical corner service station spills, the explored depths reached during hydrogeological studies are seldom intended to reach the bottom of the aquifer. Contaminants of concern at these sites are for the most part LNAPL or "floaters," and as such, aquifer thickness is not a likely priority. At sites where the contaminant is a DNAPL or where more than one aquifer is encountered, aquifer thickness will be a vital part of the hydrogeological characterization process.

7.2.1.3 Groundwater Flow Direction and Hydraulic Gradient

Groundwater flow direction depends on the hydraulic gradient. The hydraulic gradient is the change in head per unit distance in a given direction.

Both parameters are necessary for plume capture zone calculations. If there is seasonal change in groundwater flow directions, the plume capture zone calculations performed based on the groundwater elevations collected during the site characterization stage can produce misleading results.

A minimum of three separate sets of water elevations of the site, with a minimum of one month between each set of readings, should be collected to verify the direction of groundwater flow. These are usually available from quarterly monitoring reports in which data from monthly sampling events are compiled. However, if there is a significant amount of free-floating product in the subsurface, it is advisable to quickly implement the product recovery and groundwater remediation systems instead of waiting to accumulate water elevations. Immediate actions to initiate product recovery and groundwater remediation will significantly decrease the potential for plume migration. In addition, the potential for recovering greater quantities of product is higher near the onset of the spill rather than later.

7.2.1.4 Water Table Fluctuations

Water table fluctuations can play a significant role in the operation of pump and treat systems. Free-floating product on top of a rising water table may not enter the well if the screens are installed too deep. A seasonally low water table that was not accounted for may, on the other hand, affect the pumping rate of the system, creating problems such as frequent cycling of the depression pump or reduced capture zones. Historical water elevations should be obtained from local county agencies and evaluated for this purpose.

7.2.2 Geological Characterization

Geological characterization provides information on the type of materials encountered in the subsurface, and the homogeneity or heterogeneity of these materials. Geologists illustrate these formations through the use of geologic cross-sections depicting the different types of materials at various depths. Data for preparing these cross-sections are obtained from borings drilled on-site. Detailed boring logs describing the properties of the soil are prepared for every boring drilled. Soil description should include the following information:

- Approximate percentages of major and minor grain size constituents
- Color and Munsell Color
- Geologic origin
- Description of moisture content, such as dry, moist, or wet
- Any visual presence of secondary permeability
- Voids or layering
- Pertinent field observations such as odor
- Field instrument readings for contamination

- Typical field screening tools are
  - Photoionization detector (PID)
  - Flame ionization detector (FID)
  - Field gas chromatography (GC)
- Description and notation of any product smearing evidence. The depth at which the smearing is evident should be documented carefully because depth of smearing is an indication of past aquifer water level variations.

Figure 7.1 shows a typical soil boring log.

7.2.3 Extent of Contamination

A definition of the horizontal and vertical extent of contamination is one of the first priorities in performing a detailed site assessment, both from regulatory and engineering standpoints. It is necessary for, among other things, performing plume capture calculations, and determining both locations and the minimum number of groundwater recovery wells. The vertical extent is also necessary for determining the screen interval of the recovery wells.

7.2.4 Groundwater Analyses

It is strongly recommended that groundwater samples be collected and analyzed for inorganics and bacterial strains. The presence of minerals or inorganics such as dissolved iron (ferrous iron) and water hardness causing cations such as calcium and magnesium in the groundwater can create considerable problems for the treatment system. In fact, most operational down-times of remedial systems that involve groundwater pumping are caused by scaling of the unit by iron bacteria slime and calcium carbonate precipitation.

All natural waters contain some dissolved mineral matter. The commonly encountered cations in groundwater are calcium, magnesium, sodium, potassium, iron, and manganese. Major anions found in groundwater include bicarbonates, sulfates, and chlorides. Hardness in water is caused primarily by calcium and magnesium cations. Hardness can be precipitated with changes in temperature, pressure and pH. When water containing calcium and bicarbonate ions is heated, insoluble calcium carbonate is formed as shown by the following equation:

$$Ca^{2+} + 2HCO_3^- \rightarrow CaCO_3 + CO_2 + H_2O$$

In the above process, carbon dioxide is released as a gas from the bicarbonate ion when the water is heated. Consequently, part of the bicarbonate changes to carbonate, which then reacts with the calcium ion to form insoluble

AAA Construction Co.
1234 North A Street
City, State, Zip

NE 1/4 of NE 1/4 of NW 1/4 of Section 23. T 11 N, R 6 E., County, State.

Water depth/elev.:		Rig, Drill. Method:	Location:	Date:	11/28/95	Boring #:
10.4	IAD	CME-75	128' North & 137' East of the Northeast	Start:	15:00	MW-5
11.95	Stabilized	4 1/4" HSA	corner of the North building on site.	End:	15:30	

Elev: (ft)	Depth: (ft)	Group Symbol:	Description of materials:	Blow Count:	HNU: (ppm)	Remarks:
91.9	3.5	ML	SILT WITH CLAY; very dark grayish brown (10YR3/2); moist; low plasticity; firm; 5% sand	2		
91.4	4.0	SP	POORLY GRADED SAND; very pail brown (10YR7/3); moist; loose; fine to medium grained sand	3		
				7	0	N = 6
				11		N = 21
89.9	5.5			10		
89.4	6.0	SP	SAME AS ABOVE;but medium dense	4		
				5	Trace	N = 12
				6		
87.4	8.0			6		
86.9	8.5	SP	SAME AS ABOVE;but wet at 9.5; medium to course grained sand	3		
				7	Trace	Collect sample for BTEX and TRPH lab analysis
				7		N = 13
84.9	10.5			6		
84.4	11.0	SP	POORLY GRADED SAND WITH GRAVEL; yellowish brown (10YR5/4); wet; loose; medium to course grained sand; 5% fine gravel	2	0	
				4		N = 10
				5		
82.4	13.0			5		
81.9	13.5	SP	POORLY GRADED SAND; yellowish brown (10YR5/4); wet; medium dense; medium to course grained sand	3		
				6	0	N = 14
				6		
				8		
79.9	15.5		BOTTOM OF HOLE			

Figure 7.1. Typical boring log.

calcium carbonate scale. In pump and treat systems, calcium carbonate precipitation as a white chalky material is observed in air stripping towers and tray aerators. The precipitation is mitigated by an increase in groundwater pH as carbon dioxide is stripped off by the aeration process. Aeration can increase the water pH to between 8 and 9 in waters with moderate alkalinity (Sawyer et al., 1993). From water chemistry, bicarbonate ions begin converting to carbonate ions at a pH of 8.3. The heat transfer from the hot blower air to the groundwater may have a slight impact also, though this is unclear.

Reduced water pressures also result in carbonate deposits. This can occur around the well screen area. As water is drawn toward the pump in the well, the groundwater flow velocity increases with decreasing distance to the well screen. This increase in velocity is accompanied by a corresponding decrease in pressure, causing carbon dioxide gas to be driven off. As illustrated earlier, this leads to a reaction between the bicarbonate ions and the calcium and magnesium cations.

A more troublesome problem encountered in the operations and maintenance (O & M) of pump and treat systems is caused by iron bacteria. There are many strains of iron bacteria, including *Sphaerotilus*, *Gallionella*, *Pseudomonas*, and *Clonothrix*. In obtaining energy for their metabolic needs, iron bacteria oxidize dissolved ferrous iron to insoluble ferric iron, as illustrated in the following equation:

$$4Fe^{2+} + 4H^{+} + O_2 \rightarrow 4Fe^{3+} + 2H_2O$$

Iron bacteria produces a slimy material that is an accumulation of bacterial cells and deposits of iron. This mass of bacteria growth and precipitated hydrated iron (III) oxides is highly effective in plugging groundwater pumping wells and above-ground treatment units. When in a dry state, such as in an air stripping tower that has been shut down for a while, the mass becomes a hard incrustation that is difficult to remove. The original source of iron bacteria is unclear. It may already be present in the groundwater and simply multiply in population as food and energy sources are increased through pumping. In other cases, iron bacteria may be introduced into wells by drilling operations. In pneumatic pumping systems, the pulse action of the pumps may cause aeration in the vicinity of the well screen and/or along the pipings, providing oxygen for the metabolic process of the iron bacteria.

7.2.5 Other Site-Specific Characteristics

Other site-specific characteristics that should be addressed in the data collection and site characterization process include the following:

- Presence of surface water bodies or wetlands in the vicinity of the site
- A fractured-aquifer matrix
- Structures that affect groundwater and/or free floating product or DNAPL flow
- Nearby wells that might be impacted by the contamination on site or influence the natural flow patterns of the groundwater, such as irrigation wells

7.3 AQUIFER TESTING

Aquifer testing is conducted to determine the rate of pumping that the aquifer can sustain and to estimate the hydraulic conductivity or transmissivity for plume capture calculations. In some cases, hydraulic conductivity tests conducted during the site characterization process may provide sufficient data for remedial design. In other cases, a pumping test prior to remedial design may be necessary to accurately estimate the rate of groundwater pumping that is necessary to capture the plume. In general, it is preferable to design a pump and treat system based on data from a pump test rather than from calculations based on hydraulic conductivity values. This is because data from an actual pump test is more reliable and will more closely resemble actual operating conditions.

For an unconfined aquifer, the pumping rate is given by the following equation:

$$h^2 = \frac{Q}{\pi K} \ln\left(\frac{r}{r_w}\right) + h_w^2 \tag{7.1}$$

where h = static head (ft)
Q = pumping rate (gpm)
K = hydraulic conductivity (gpd/ft^2)
r = radius of influence (ft)
r_w = well radius (ft)
h_w = water level in the well during pumping (ft)

If static head readings are plotted versus distances on semi-log paper, the slope can be used to estimate the hydraulic conductivity. A least squares analysis using all of the field data points can also be used to calculate the slope, which may be used to estimate the hydraulic conductivity.

For a confined aquifer, the pumping rate is given by the following equation:

$$s = \frac{Q}{2\pi T} \ln r - \frac{Q}{2\pi T} \ln r_w \tag{7.2}$$

where s = drawdown (ft)
Q = pumping rate (gpm)
T = transmissivity (gpd/ft)
r = radius of influence (ft)
r_w = well radius (ft)

If drawdowns are plotted versus distances on semi-log paper, the slope can be used to estimate the transmissivity. Again, a least squares fit can be used to estimate the slope.

Sometimes, simple aquifer testing techniques such as slug tests, bail down tests, and grain size analysis methods can be used to provide sufficiently accurate hydraulic conductivity estimates. However, these techniques may not be sufficiently accurate for design purposes.

The following is a list of aquifer tests in decreasing order of accuracy:

- Long duration (more than one day) constant-rate pumping tests
- Short duration (less than eight hours) step drawdown tests
- Bail down slug tests
- Permeability calculations based on grain size analysis

Following are some general rules of thumb for deciding what types of aquifer tests are sufficient:

- In sand or gravel formations, an aquifer test is required if the contaminant plume is hundreds of feet long and over 100 feet wide.
- In silt and clay soils, a bail down slug test is generally sufficient for evaluating remediation alternatives.
- For a pump and treat system that is capable of producing more than 50 gpm, an aquifer test is needed.
- For a pump and treat system that is not likely to exceed 5 gpm, an aquifer test is not needed.
- If the system is expected to operate between 5 to 50 gpm, other factors such as cleanup levels, water disposal options, etc., need to be assessed for determining what level of accuracy is necessary for an aquifer test.

7.3.1 Pump Tests

A constant-rate pump test extracts groundwater at a constant rate for a number of hours, and a step drawdown test varies the pumping rate over time to verify the capacity of an aquifer to yield water. A pump test is the best method for calculating aquifer transmissivity and specific yield or storage coefficient. A pump test can be performed in an aquifer test well constructed for the pump test, a groundwater recovery well, or an oversized monitoring well.

Disposal of contaminated groundwater recovered from a pump test is a consideration in most cases. In some jurisdictions, this may be a more sensitive issue than others. If the pump test well is conducted upgradient of the source and within the same geological unit, it may produce clean water. Disposing of clean water from a pump test is much easier and cheaper than contaminated water.

Typically, a pump test for an unconfined aquifer is run for 72 hours, while confined aquifers may require only a 24-hour test. At some small sites, a low capacity test (less than 10 gpm) for a shorter period of time (8 to 24 hours) may be sufficient.

During the pumping test, water level measurements should be collected at all available measuring points. Even distant points that are outside the radius of influence can provide data on background water level fluctuations during the test. In any case, all recovery data from a pumping test should be evaluated, especially at the groundwater recovery well.

7.3.2 Bail Down and Slug Tests

A bail down test is a test that instantaneously extracts a volume of water or a slug from the well and measures the return to static water levels as a function of time. A slug test, on the other hand, is a test in which a solid slug is instantaneously injected into the well and the return to static water levels is measured as a function of time. Both bail down and slug tests provide poorer hydraulic conductivity estimates than pump tests but better estimates than grain size analyses.

Both bail down and slug tests are particularly suited for determining aquifer characteristics of low permeability materials. This is because materials with low permeability ensure a relatively slow rate of water level recovery, which facilitates accurate measurements in the well. In addition, they are convenient and inexpensive to perform since no pumping equipment is required.

Bail down and slug tests are less accurate than pump tests because they only evaluate that part of the aquifer immediately adjacent to the filter pack and screen. If there is flow in secondary porosity channels, the well may not intersect the channels or fractures and would only evaluate the primary permeability. In addition, smearing of the borehole during drilling will cause the well to reflect an artificially low permeability.

High permeable aquifers often yield artificially low estimates with bail down/slug tests because the extraction/injection rate relative to the rate of the induced outflow/inflow from the aquifer is not instantaneous. If the filter pack is less permeable than the native soil, the calculated hydraulic conductivity is artificially low because the test measures the hydraulic conductivity of the filter pack. A screen slot size that is too small can also limit the groundwater flow into a well, lowering the hydraulic conductivity estimate in high permeable aquifers.

7.3.3 Hydraulic Conductivity Estimates Based on Grain Size Analysis

Because hydraulic conductivity depends on porosity, grain size and distribution, and the continuity of pores, grain size analysis can be used as a rough estimate of hydraulic conductivity values. Hydraulic conductivity estimation based on the grain size analysis, however, is rarely appropriate for designing a pump and treat system.

7.4 SYSTEM DESIGN

As in the design of SVE/AS and bioremediation systems, the design and installation of pump and treat systems is best served by following the systematic approach described in Chapter 4. After setting up the design folder or binder, the first step is compilation of all site characterization data. It is always good practice to neatly document all previous findings and conclusions. Sometimes this effort may seem redundant, especially if the engineer was involved in the data collection and site characterization process and is extremely familiar with the site. However, there are several advantages to spending the extra effort to perform this task: (1) The findings and the conclusions of the site characterization and pump tests provide the platform from which the basis of the final design is developed. Thus, a short summary of the data collected and results of all tests performed is a vital part of the design report. (2) It is highly likely that the conceptual design will be reviewed by a third party, whether it be the client or a regulator. (3) Projects are often transferred from one engineer to another (sometimes in different firms), and oftentimes inadequately documented design information is lost in the process.

After reviewing and analyzing the site characterization data and pump test results, the objectives of the remedial system can be defined. These may involve the design and installation of a system for recovery of free-floating product. Or the pollutant may be a DNAPL and the extent of contamination so widespread that product recovery is not economically feasible. Then the objectives shift to the design and installation of a system for the purpose of achieving hydraulic control over the migrating plume. Regardless of the technology that will be utilized, the engineer must always approach his or her design well aware of exactly what the system strives to achieve.

The next step is to establish design criteria that would form the basis for the design process. These parameters and assumptions will be used in all design calculations and treatment equipment and materials selection. Because it involves the movement of groundwater to above-ground treatment systems, pump and treat is a more complex system to design than an SVE/AS system. There are more variables to consider and more alternatives to analyze. Groundwater recovery and product recovery systems may consist of a single high capacity well, multiple low capacity wells, or a shallow trench system. It may be appropriate to install a groundwater recovery or product recovery well in the source area to minimize free-floating product migration and maximize product recovery. It may also be necessary to install a groundwater recovery well further downgradient to capture dissolved phase contaminants. It is obvious that no specific recovery system design is appropriate for all conditions; a system should be tailored to meet site-specific conditions, contaminant characteristics, and design objectives.

The next step is the detailed design of the remedial system. In pump and treat, it typically involves capture zone determination, groundwater recovery well design, pump selection, above-ground treatment unit design, bacterial growth and mineral deposits treatment, miscellaneous equipment and materials selection, electrical power availability and requirements, and treated groundwater disposal.

The next step involves the development of an O & M plan for maintenance of the system to ensure optimal performance. It does not matter how well the treatment concept is thought out and constructed if the system is shut down more than 10% of the time. This is why it is important to include O & M considerations in the fourth step and perform the two concurrently. The design of the system should take into account O & M factors such as:

- Layout that permits easy access and adequate workspace
- Instruments that are easily removed and repaired or replaced
- Potential carbonate scaling of system
- Likelihood of iron bacterial growth problem

7.4.1 Capture Zone

Groundwater recovery systems are designed to contain and remove contaminated groundwater from the aquifer. The size of the plume which the recovery system will be designed to recover varies from site to site depending upon factors such as aquifer conditions, physical and chemical characteristics of the contaminants, degree of contamination, distribution of contamination, and the location of receptors.

A groundwater recovery system may be operated as source control, as aquifer restoration, or for both purposes. If large amounts of free product are present on-site, the system may consist of two recovery wells: one for free product recovery in the source area, and one downgradient to capture the dissolved plume. If only a small amount of free product is present, a single well with total fluid recovery can be used. In this case, the recovered free product and groundwater will be pumped into an oil-water separator where the free product is separated from the groundwater. The groundwater is then pumped into an air stripper for treatment.

The capture zone of the recovery well should not "undershoot" the area of contamination and allow contamination to escape or "overshoot" the area of contamination and recover clean water that does not need treatment. However, an overshoot capture zone, over and above the zone of contamination, is sometimes warranted if there is a low level of confidence in the distribution of contamination or the aquifer testing results.

The capture zone can be calculated from the following equation:

$$x = \frac{-y}{\tan\left[\frac{2\pi Kbiy}{Q}\right]} \tag{7.3}$$

where K = hydraulic conductivity, ft/day
b = aquifer thickness, ft
i = prepumping hydraulic gradient, ft/ft
Q = pumping rate, ft^3/day

The distance to the downgradient stagnation point created by the pumping well is given by:

$$x_0 = \frac{-Q}{2\pi Kbi} \tag{7.4}$$

The maximum half-width of the capture zone as x approaches infinity is given by:

$$y_{max} = \pm \frac{Q}{2Kbi} \tag{7.5}$$

It can be seen from these equations that the distance to the downgradient stagnation point is propotional to the pumping rate but inversely proportional to the gradient and hydraulic conductivity. Thus, the greater the pumping rate or the smaller the hydralic gradient, the greater the distance to the downgradient stagnation point. In addition, they also show that the width of the capture zone is proportional to the pumping rate but inversely proportional to the hydraulic gradient and hydraulic conductivity. Thus, the greater the pumping rate or the smaller the hydraulic gradient, the wider the capture zone.

Example: Given the following information, calculate the capture zone, downgradient stagnation point, and the total width of the capture zone.

Q = 50,000 ft^3/day
b = 30 ft
K = 1,000 ft/day
i = 0.006

The capture zone is calculated to be

$$x = \frac{-y}{\tan\left[\frac{(2)(3.1415)(1,000 \text{ ft/day})(30 \text{ ft})(0.0006)y}{10,000 \text{ ft}^3/\text{day}}\right]} \tag{7.6}$$

The capture zone is plotted in Figure 7.2 with a set of predetermined y values.

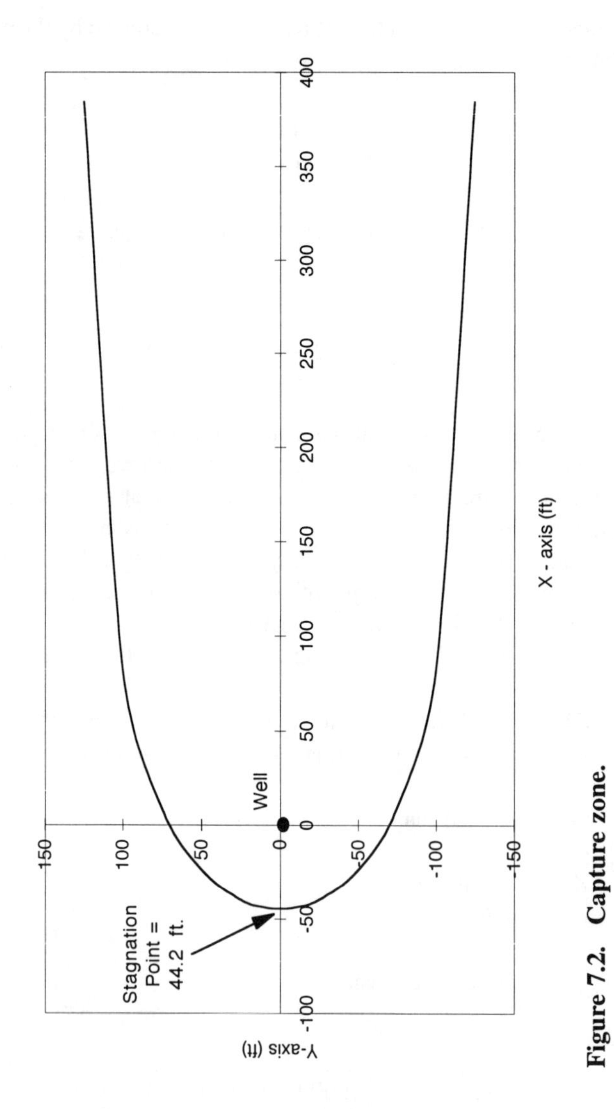

Figure 7.2. Capture zone.

The distance to the downgradient stagnation point is calculated as:

$$x_0 = \frac{-50,000 \text{ ft}^3/\text{day}}{(2)(3.145)(1,000 \text{ ft/day})(30 \text{ ft})(0.006)} = -44.2 \text{ ft}$$

and the total width of the capture zone upgradient of the well is calculated as:

$$\text{Total } y_{max} = \frac{(2)\left(50,000 \text{ ft}^3/\text{day}\right)}{(2)(1,000 \text{ ft/day})(30 \text{ ft})(0.006)} = 277 \text{ ft}$$

If a single well does not provide the capacity necessary for plume capture, several alternatives are available to assure that the recovery system will deliver the desired performance. These include:

- Multiple recovery wells can be used with reduced flow rates, which reduces drawdown in each well. When multiple wells are used, superposition can be used to estimate the drawdown in each well.
- Pumping from a trench system instead of from a recovery well.
- The length of the screen can be increased. If this option is considered, the costs associated with pumping, treating, and disposing of clean water that is pumped from under the plume should be evaluated. This option may seem cost-effective initially because it moves more water at a minimal cost. But, in some cases, the treatment and disposal costs for pumping clean water for many years make this a high-cost option.

After installation and startup of the groundwater recovery system, water table maps should be prepared periodically to depict the capture zone to ensure the proper operation of the system.

7.4.2 Groundwater Modeling

For sites with multiple extraction wells, groundwater flow calculations become very complex and it is difficult to perform hand calculations. With the increased utilization of desktop computers, more and more computer programs are written to model groundwater flow. Thus, groundwater modeling via computers can be performed to quickly and inexpensively evaluate groundwater flow to extraction wells.

The groundwater modeling programs allow the user to model a groundwater recovery system under differing natural gradients to assure that the extraction wells are in optimal locations and have sufficient pumping rates to remediate the site. There are many groundwater models available, a few of the popular ones are MODFLOW, RANDOM WALK, USGS SUTRA, and SWIFT.

MODFLOW is perhaps the most widely used groundwater model. The model is three dimensional, employs the finite difference technique, and takes into account unconfined, confined, or mixed aquifer systems. It enables the user to realistically simulate both aquifer recharge and aquifer discharge.

RANDOM WALK is a popular, user friendly, finite difference model used to simulate either one-dimensional or two-dimensional steady or unsteady flow, and solute transport in nonhomogeneous, unconfined, or confined aquifers.

USGS SUTRA was developed under a joint project between the USGS and US Air Force. It is a finite element code that may be used to simulate two-dimensional, saturated/unsaturated, density-dependent flow, heat, mass transport.

SWIFT is a complex, but highly flexible finite difference model that can be used for the simulation of coupled density-dependent flow and transport of heat and contaminants, in both porous and fractured media.

7.4.3 Groundwater Recovery Well Design and Construction

A groundwater recovery system in high-permeable soils typically consists of a high capacity well or multiple high capacity wells depending on the size of the contamination plume. On the other hand, a groundwater recovery system in low-permeable soils typically consists of a trench system or multiple low capacity wells. The factors deciding the utilization of multiple low capacity wells instead of a trench system include (1) a trench system may be difficult or impossible to install because the trench does not stay open long enough for pipe installation and backfilling especially a deep trench, and (2) installing a deep trench may be cost prohibitive.

There are four basic elements to the design and construction of groundwater recovery wells: the casing, the grout or seal, the intake or screen portion, and the filter or gravel pack. The casing serves to connect the aquifer with the surface and also as a housing for the pumping equipment. It allows access to the aquifer and supports the sides of the borehole. The grout or seal is the material used to seal the original borehole in the annulus outside of the casing. It serves to prevent vertical migration of water along the sides of the casing and borehole. The propose of the intake or screen is to allow groundwater to enter the well while preventing the migration of sediments into the well. In addition, it provides a stabilizing structure for supporting the loose aquifer material. Filter or gravel packs are placed in the space between the screen portion and the borehole.

7.4.3.1 Well Diameter

In pump and treat systems, it is typical to construct the casing and screen portion of the well with the same diameter. Well yields are affected by screen diameter, although doubling the diameter will only increase the yield by approximately 10%. The equation for the well yield of an unconfined aquifer is

$$Q = \frac{K(H^2 - h^2)}{1{,}055 \log(R/r)} \tag{7.7}$$

where Q = well yield or pumping rate (gpm)
K = hydraulic conductivity (gpd/ft^2)
H = static head (ft)
h = depth of water in the well during pumping (ft)
R = radius of the cone of depression (ft)
r = radius of the well (ft)

The equation for a well pumping under confined aquifer conditions is

$$Q = \frac{Kb(H - h)}{528 \log(R/r)} \tag{7.8}$$

where b = thickness of aquifer (ft)

In recovery wells, the casing diameter must be large enough to accommodate the pump and its appurtenances, and in some cases, the product pump. It must also provide sufficient room for ease of maneuverability when taking water level measurements and performing other O & M activities. A separate 2-inch slotted polyvinyl chloride (PVC) pipe could be placed within the well to serve as a conduit for dropping a water level tape or product interface tape. This helps prevent the tape from getting entangled with the discharge, electrical, and sensor probe lines.

7.4.3.2 Well Depth and Well Screen Length

The well depth and optimum length of well screen is a function of aquifer characteristics, vertical extent of contamination, and designed yield.

7.4.3.3 Well Casing and Screen Materials

Once the well diameter and length has been determined, the next step in the well design process is selection of the type of casing and screen material. There are five major types of well casing and screen materials. They are PVC, polypropylene (PP), teflon, mild steel, and stainless steel. All of the plastic materials are lightweight and chemically resistant to alkalies, acids, hydrocarbons, and oils. In terms of cost, teflon is much more expensive than PVC and

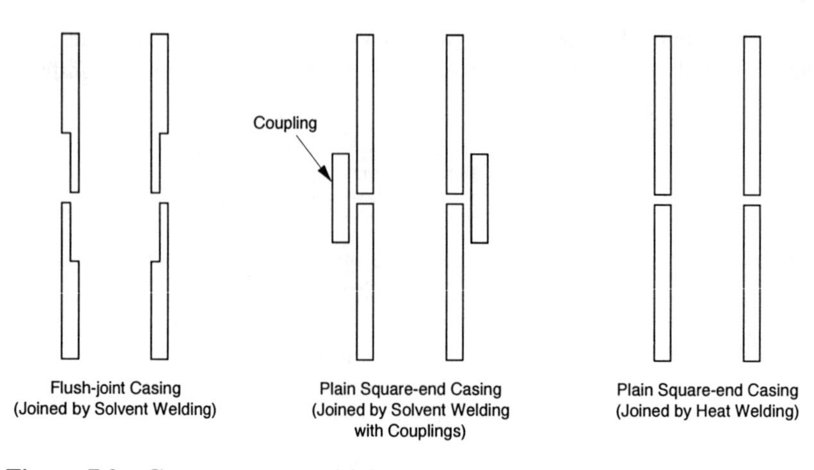

Figure 7.3. Common types of joints.

PP. Generally, the plastic materials are weaker, less rigid, chemically more resistant, and more temperature sensitive than metallic materials.

7.4.3.4 Well Screen Type

Three types of well screen are commonly used. They are field slotted pipe, factory slotted pipe, and wire-wound continuous slot screen. Both field slotted and factory slotted pipes are relatively inexpensive compared to the wire-wound continuous slot screen. However, they generally have slots that are too large, which will cause an excessive amount of material to enter the well. Wire-wound continuous slot screen has very good slot control with a wide range of slots available.

7.4.3.5 Joint Types

There are only a limited number of methods available for joining lengths of casing or casing and screen together. The joining method depends on the type of casing and type of casing joint. Flush joint, threaded flush joint, plain square end, and bell end casing joints are typical of joints available for plastic casing. Threaded flush joint, bell end, and plain square end casing joints are typical of joints available for metallic casing. Figure 7.3 shows some common type of joints available.

7.4.3.6 Filter Pack

The term filter pack or gravel pack refers to a coarse-graded material that is introduced into the annular space between a centrally positioned well and the borehole. The primary purpose of the filter pack is to retain the formation material and keep the water entering the well as sediment free as possible. Because the filter pack is more permeable than the surrounding formation, it increases the effective hydraulic diameter of the well.

7.4.3.7 Well Screen Slot Size

For artificially filter-packed wells, the well screen slot size is generally chosen on the basis of its ability to hold back between 85 to 100% of the filter pack material. For naturally packed wells, well screen slot sizes are generally selected based on the following criteria:

1. The slot size should be able to retain no less than 30% of the formation material if the uniformity coefficient of the formation material is greater than 6 and the material above the intended screened interval is noncaving.
2. The slot size should be able to retain no less than 50% of the formation material if the uniformity coefficient of the formation material is greater than 6 and the material above the intended screened interval is readily caving.
3. The slot size should be able to retain no less than 40% of the formation material if the uniformity coefficient of the formation material is less than 3 and the material above the intended screened interval is noncaving.
4. The slot size should be able to retain no less than 60% of the formation material if the uniformity coefficient of the formation material is less than 3 and the material above the intended screened interval is readily caving.

7.4.3.8 Grouting Materials

The annular space between the well casing and the borehole needs to be sealed at some point to prevent infiltration of surface water and potential contaminants from the ground surface. The two most effective materials for this purpose are bentonite and neat cement.

Bentonite is a hydrous aluminum silicate comprised principally of the clay mineral montmorillonite, while neat cement is a mixture of Portland cement and water in the proportion of 5 to 6 gallons of clean water per 1 cubic foot of cement.

7.4.3.9 Drilling Methods

The five well drilling methods that are commonly used to drill groundwater recovery well are described below.

Hollow Stem Auger Method

The hollow stem auger method is a fast and effective means of drilling and completing small diameter wells to moderate depth. With this method, borehole caving and sloughing can be prevented by placing screen and casing in augers before the augers are removed. The common diameter of wells drilled with the hollow stem method range from 6.25 to 13 in. No drilling fluid is needed with this drilling method, which minimizes contamination problems. In addition, formation water can be sampled during drilling by using a screened lead auger.

The major disadvantages of the hollow stem auger drilling method are that it can be used only in unconsolidated materials and it is limited to a depth of 100 to 150 ft.

Solid Stem Auger Method

The solid stem auger method is an effective drilling and relatively low-cost method in completing shallow and small diameter wells of up to 14 in. in diameter. Sample recovery is excellent with this drilling method in unsaturated conditions. Undisturbed soil samples can be taken by removing the augers and sampling in the open hole.

The major disadvantages of the solid stem auger drilling method are that sample collection below saturation zone is poor because an open hole is not maintained by the augers, vertical aquifer sampling is impractical, and it is limited to a depth of less than 200 ft.

Mud Rotary Method

The mud rotary drilling method is a relatively fast and inexpensive drilling method that can be used in both unconsolidated and consolidated formations. It is capable of drilling to any depth while allowing the collection of core samples during drilling. Its major disadvantage is the requirement of drilling fluids that will mix with formation water and invade the formation that is usually difficult to remove. Also, no information on the location of the water table and only limited information on water-producing zones is directly available during drilling.

Air Rotary Method

The air rotary method is another drilling method that does not require drilling fluids and can be used in both unconsolidated and consolidated formations. It is capable of drilling to any depth. With this drilling method, formation water is blown out of the borehole along with cuttings, making it possible to determine when the first water-bearing zone is encountered. The collection and analysis of water blown out of the borehole can provide information regarding changes in water quality.

This method is relatively expensive, thus it may not be economical for small jobs. In addition, casings are required to keep the borehole open when drilling in soft, caving formations, such as below the water table.

Cable Tool Method

The cable tool method is the oldest drilling method that encompasses a wide depth range. It can be used in both unconsolidated and consolidated formations, but it is most suitable for caving large gravel-type formations with large cavities above the water table.

This drilling method is a slow drilling method and it requires a small amount of drilling fluid that will create contamination problems. With this method, relatively large diameters are required and screens must be set before a water sample can be taken. In addition, there is a potential difficulty in pulling back the casing when the drilling is completed.

7.4.3.10 Well Development

After construction is completed, the well should be developed to correct damage done to the borehole during drilling. Well development will remove clays, silts, or fine sands from the formation adjacent to the well screen to minimize or eliminate the pumping of fine particles, and to stabilize the borehole. Well development can be accomplished with overpumping, surging, and water jetting.

Overpumping involves pumping the well at a rate substantially higher than it will normally be pumped. The intent is to pump the well at the highest rate attainable, increasing the drawdown in the well to the lowest permissible level. This results in increased flow velocities that will induce the flow of silts, clays, and other debris into the well, thus opening screen slots and pore spaces, and/or cleaning fractures.

Surging involves the rapid dropping of a surge block into the well, forcing water contained in the borehole into the aquifer surrounding the well. In withdrawing the block, the water is lifted by the surge block, allowing the flow of water and fine sediments back into the well from the aquifer.

Water jetting can open fractures and remove drilling mud that has penetrated the aquifer. Water jetting is accomplished by injecting water into the well through nozzles on a jetting tool.

7.4.4 Trench System

Trench systems are only used in groundwater recovery systems if the water table is very shallow and the soil is of low permeability. Trench systems are typically installed with a backhoe. The purpose of the trench is to create a high-permeable channel through the native soil to recover more groundwater than a well possibly could. The saturated zone of the trench should be backfilled with a high-permeable material, such as coarse sand or gravel. If the trench is very long, a perforated pipe or well screen should be installed horizontally in the base of the trench to conduct water to a recovery well or sump.

The unsaturated zone of the trench can be backfilled with the noncontaminated soils that were originally excavated from the trench. In some cases, a geotextile can be installed above the coarse gravel and below the backfill. If free product is present, the coarse backfill should extend one or more feet above the seasonal high water table to ensure that free product will not rise into the native fine-grained backfill.

7.4.5 Hydraulics and Pump Selection

In designing a groundwater remedial system, the design engineer will be required to select a pump(s) for pumping the contaminated groundwater to the surface for treatment. To that end, an understanding of basic hydraulics and pump selection is essential.

7.4.5.1 Groundwater Recovery Pumps

For pumping groundwater, there are typically two types of pumps — those installed on the ground surface and those installed in wells. Surface-mounted pumps are generally limited to wells with high static water levels and small drawdowns, since the maximum practical suction lift is limited to about 20 ft. Both constant-displacement and variable-displacement pumps can be used as surface pumps, with the most commonly installed surface pump being the variable-displacement centrifugal pump.

Both constant-displacement and variable-displacement pumps can be installed in wells. There are two types of vertical centrifugal pumps, depending on the location of the prime mover: a deep-well vertical turbine pump driven by a rotating shaft connected to an electric motor at the surface, and a submersible pump driven by a motor connected directly below the pump. The advantage of the submersible pump is the elimination of the long drive shaft and bearing assemblies connecting the pump drive at the surface to the bowl assembly near the intake. However, a submersible motor can be a problem since it normally operates at higher speeds and is not directly accessible for observation and repair.

Pneumatic pumps are also commonly used in pump and treat systems. As a general rule, electric submersible pumps are used when the designed flow rate is greater than 5 gpm. For lower flowrate applications (less than 5 gpm), both submersible pneumatic and electric pumps can be used. Pumps should be constructed of materials that are compatible with the contaminants present at the site. Similarly, motor leads, seals, and bearings should also be made of materials that are compatible with the site contaminants.

In general, submersible pumps do not have to be explosion proof because the pump motor is below the intake of the pump, therefore the pump motor will always be submerged and isolated from contaminant vapors. Electric sump pumps that have a motor above the pump inlet should be explosion proof.

Where possible, avoid placing the pump intake of a submersible electric pump within the screen interval of the well. This is because the flow patterns created will greatly increase the groundwater flow velocities in the vicinity of the screen causing deposits as discussed earlier. Increased velocities can also cause sand pumping.

In selecting a pump, the engineer must give the pump suppliers as much data and general information as possible. Typically, this data will be included in the project technical specifications for bidding purposes. A general outline of information that the engineer will need to determine or provide are

- Capacity (flowrate)
- Total dynamic head
- Net positive suction head available
- System head curve
- Horsepower characteristics
- Drives, electric motors
- Cost of pumping

Capacity

The capacity of the aquifer is usually known from hydrology data and pump tests.

Total Dynamic Head

The hydraulic calculations begin at a control point where the system discharges to a receiving water course or sewer (gravity flow under atmospheric conditions) or pressure tank or forced main (pressurized flow). The system's pumping requirements are established from this control point.

Net Positive Suction Head

Net positive suction head (NPSH) is defined as the total amount of energy, in absolute feet of water, at the pump centerline. There are two types of NPSH: available and required. Net positive suction head required (NPSHR) is the amount of absolute pressure needed to get the water into the pump and keep the pump running efficiently, taking into consideration head losses due to friction within the pump mechanisms. NPSHR is a characteristic of the pump and is provided by the manufacturer (generally as part of the pump curve). Net positive suction head available (NPSHA) is the energy actually available, in feet of absolute pressure, at the inlet of the pump. It depends on the location and design of the intake system and can be calculated by the engineer.

$$NPSHA = H_{abs} + H_s - H_f - H_{vp} \tag{7.9}$$

where NPSHA = net positive suction head available (ft)
H_{abs} = absolute pressure on the surface of the liquid in the suction well (ft)
H_s = static elevation of the liquid above the centerline of the pump (ft)
H_f = friction head and entrance losses in the suction piping (ft)
H_{vp} = absolute vapor pressure of fluid at the pumping temperature (ft)

This is illustrated graphically in Figure 7.4. It is essential that the NPSHA be greater than NPSHR with a factor of safety of approximately 2 to 3 ft. When the NPSHR is equal to or greater than the NPSHA, cavitation can occur. Cavitation refers to the formation of gas bubbles (vaporization) at points of low pressure, such as at the pump inlet, and their eventual collapse under high pressure. From Bernoulli's thereom, when velocity increases, such as at the impeller, there is a decrease in pressure. Under reduced pressure conditions (or similarly, at elevated temperatures), the waters' tendency toward vaporization increases.

At higher elevations, water boils at lower temperatures. (At 212°F, water has a vapor pressure of 14.7 psia, the same as atmospheric pressure, and it boils.) Thus, if the flow is subjected to an ambient pressure equal to or less than the vapor pressure of the water, gas bubbles will form. Collapse of the bubbles (implosion) takes place when they move into regions where the local pressure is greater than the water vapor pressure. This collapse occurs with considerable force against the impeller vane, chipping and disintegrating the metal bit by bit.

System Head Curve

When selecting pumps for a specific application, the design engineer must match the pump's performance with the head capacity curve for the system. The total system head consists of:

- Static head, which is the difference between the elevation of the highest point to be pumped to and the dynamic water level.
- Major head loss, which is pressure drop due to pipe friction.
- Minor head loss, which is head loss from all other fittings and devices such as flow meters, valves, elbows, and changes in pipe sizes.

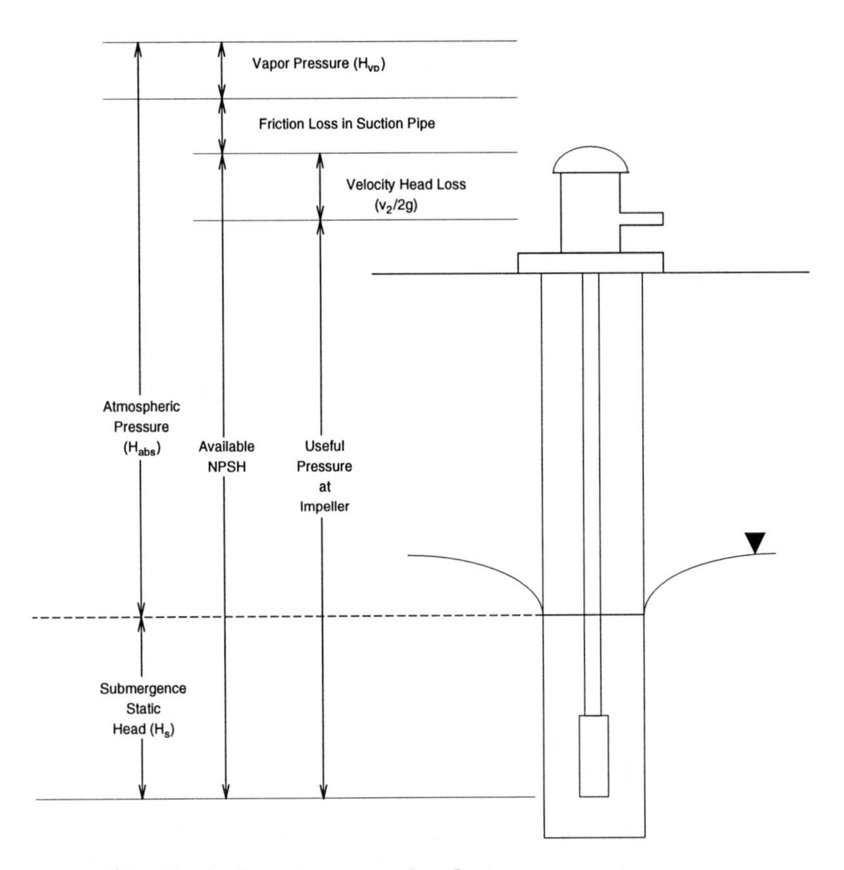

Figure 7.4. Typical pump system head curve.

Figure 7.5 is an illustration of a typical system head curve. Pumps are selected based on the designed pumping rate and the total system head to be overcome. If pressure is required for flow across the treatment unit, then the total system head must include the pressure drop across the unit. Given the pipe diameter and flowrate, head loss due to friction can be computed from the Hazen-Williams nomograph. It can also be calculated from the Darcy-Weisbach equation,

$$H_f = f\left(\frac{L}{d}\right)\left(\frac{v^2}{2g}\right) \tag{7.10}$$

where H_f = head loss due to friction (ft)
f = friction factor, which is a function of flow type (usually laminar) and pipe roughness
L = pipe length (ft)
d = pipe diameter (ft)
v = flow velocity (ft/sec)
g = gravity acceleration (ft/sec²)

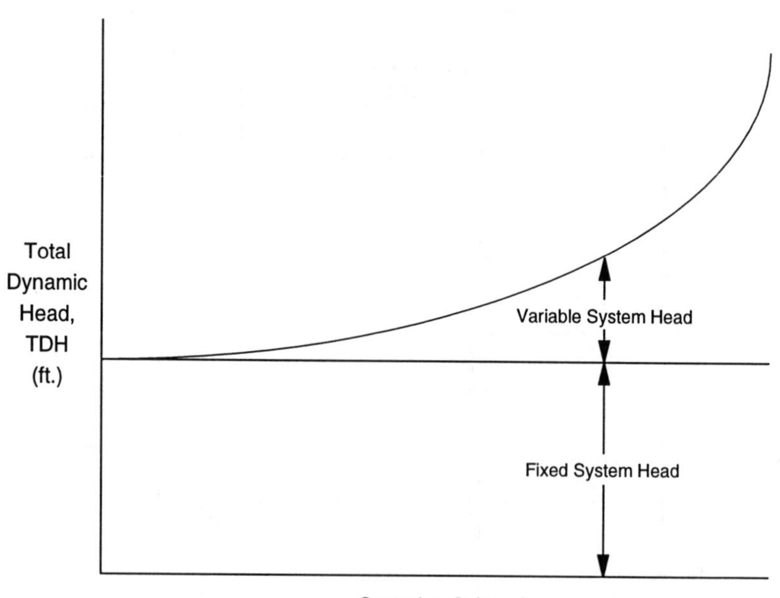

Figure 7.5. NPSH vertical turbine pump.

Minor head losses are usually only a small fraction of the major head loss. But in a relatively short length of pipe with a significant number of fittings, instruments, and turns, the total losses may add up to a fairly significant amount. Minor head losses for various devices and fittings are easily calculated using a nomograph that converts each fitting to an equivalent pipe length. Alternately, minor head losses can be computed by multiplying the velocity head ($v^2/2g$) by the respective coefficient for each condition. All the nomographs and coefficients described in this section can easily be found in most textbooks dealing with hydraulics and water supply design.

Horsepower Characteristics

Pump motor horsepower is determined by the following formula:

$$\text{Motor HP} = \frac{\text{Flow (gpm)} \times \text{Total System Head (ft)}}{3{,}960\ E_{pm}} \tag{7.11}$$

E_{pm} is the efficiency of both the pump and the motor and is called the wire to water efficiency. It is obtained by multiplying the pump efficiency by the motor efficiency. Usual ranges for pump and motor efficiencies are 50 to 85% and 80 to 95%, respectively.

Valves

Along with a flowmeter with totalizer, a throttle valve is added to the line near the treatment system to artificially create more head to obtain the desired flowrate. However, if a throttle valve is used, care should be exercised to avoid burning out the pump by creating too much restriction to groundwater flow. A pressure gauge marked with the maximum pressure may also be installed in the line near the throttle valve to prevent accidental damage to the pump.

Drives

Electrical connections to the pumps must be designed to meet specifications that are acceptable to the local electrical inspector. The wire insulation to the pump motor should be compatible with the site contaminants. If the contaminants are ignitable, the local electrical inspector may require an explosion-proof junction box.

Compressed air lines that are used for pneumatic pumps can freeze up in cold weather, and should be protected from subfreezing conditions. The most common method of preventing the air lines from freezing is by burying them beneath the frost line (approximately 5 feet below ground surface). It is also important to ensure that there are no kinks in the air lines. Water traps are recommended for air compressors with air tanks for the purpose of draining condensate.

Cost of Pumping

The cost of pump operations can be calculated for an electric motor as:

$$\text{Cost per Hour} = \frac{\text{Flow (gpm)} \times \text{Total System Head (ft)} \times 0.746 \times \text{kW} - \text{Hr (Cent)}}{3{,}960 \times E_{pm}} \quad (7.12)$$

7.4.5.2 Free-Floating Product and DNAPL Pumping Systems

Pumps installed in recovery wells to pump free-floating product or LNAPL are called recovery or product pumps. These pumps may operate pneumatically or may be electric submersible pumps. Product pumps may operate as part of a two-pump system (in conjunction with a groundwater depression pump) or independently (operating as just a product skimmer).

In a two-pump system, product pumps remove the product that accumulates into the well as a result of the cone of depression created around the well by the depression or deep pump. The product pump can be either a floating type or nonfloating type. Pumps with in-built float mechanisms can automatically adjust to changing water levels in the well; however, they typically operate within a limited range (approximately two feet). Because dynamic water levels in the well can fluctuate unpredictably (due to a variety of reasons, including pumping rate variations, well recharge changes, plugging of well screens, increased head, etc.), floating pumps may need to be adjusted frequently.

Another type of recovery pump deploys a filter that allows product, but not water, to flow in through the pump inlet. Operational problems with this type of pump include clogging of the filter with fine materials present in the well. In addition, the pumping of groundwater tends to create an emulsion that exists between the groundwater and free product interface. This emulsion in turn can "confuse" the filtration mechanism and cause the recovery pump to pump a mixture of water and product.

Pumps that operate on preset elevations of product levels can be difficult to use because of fluctuating elevations. Typically, to facilitate regular pumping of product accumulating in the well, both "on" and "off" sensors for the product recovery pump have to be adjusted on a regular basis. Where the environment in the well is harsh from a contaminant, biological, and chemical standpoint (which is normally the case), the sensors may have to be cleaned regularly.

7.4.5.3 Total Fluids Pumps

Sites with very low permeability, and thus low yielding wells, often use total fluid pneumatic pumps. A total fluid pump pumps all fluids — floating product, water, DNAPL. The pumped liquid is usually discharged into a product separator prior to the treatment system. Floating product is recovered from the top of the separator tank and the recovered water is either treated prior to discharge or drained to the sanitary sewer (if concentrations and/or flowrates are low enough to be acceptable to the local POTW). Pneumatic pumps are often used in low-capacity applications because they can safely run dry without danger of burning out or suffering damage from running dry.

Because of their mechanism of operation, total fluids pneumatic pumps are inherently more subject to biofouling than other types of pumps. The oxygen-enriched water in the discharge lines of pneumatic pumps provides the perfect environment for iron bacteria; the longer the lines, the greater the amount of biofouling. This problem extends beyond the pump and discharge lines; separator tanks are subject to the same amount of biofouling.

7.4.7 Air Stripping Tower Design

Air stripping is a mass transfer process that is effective in removing VOC. The air stripping process relies on the simple fact that volatile organic compounds (VOC) evaporate more readily than water. In a stripping tower, air and water are run countercurrent through randomly packed structure media. Contaminated water is introduced into the tower through a distributor located near the top of the tower, whereas air from a blower is forced upwards into the tower. The contaminated water trickles through the stack of packing that enhances air/water contact by breaking the water into thin films and exposing a large amount of liquid surface area to the upward flow of air. The VOC evaporate and are carried away by the upward flowing air to the atmosphere or vapor treatment units, such as activated carbon absorption units and thermal destruction units. The clean water that has been stripped of VOC drops to the sump of the tower and subsequently discharges into surface water bodies, sanitary sewers or reinjected into the subsurface. Figure 7.6 depicts a typical air stripping tower system.

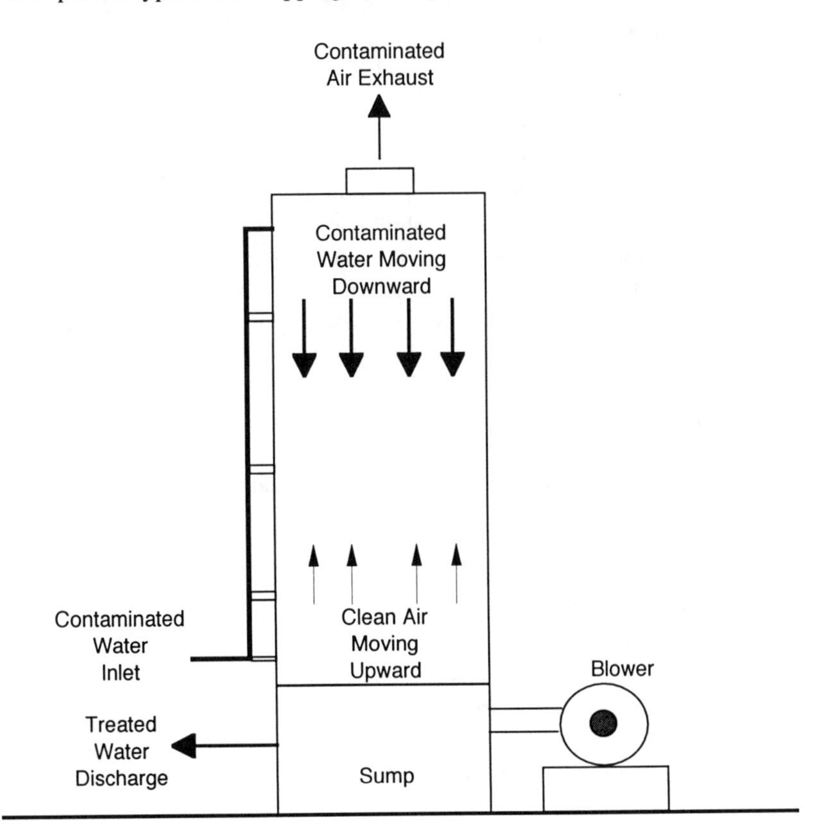

Figure 7.6. Typical air stripping tower set-up.

The VOC removal efficiency of an air stripping tower depends greatly on the following factors:

- Water temperature
- Air to water ratio
- Tower height and diameter

For most VOC, air stripping at elevated temperatures will increase removal rates. Additionally, increasing the water temperature will make some normally difficult to strip VOC like acetone be removed more effectively. Increasing the temperature to improve removal efficiency is not a feasible idea because of the huge energy cost involved. Besides, where water hardness is a consideration, the precipitation of calcium carbonate will create an O & M nightmare. Even without temperature elevations, a reduction in pH due to the simultaneous stripping of carbon dioxide may cause precipitation problems.

The required air to water ratio of an air stripping tower depends on the volatility of the contaminants. The more volatile a contaminant, the smaller the air to water ratio is needed to strip the contaminant off the water stream. Air to water ratios typically range from 10:1 to 200:1 depending on the compound being removed.

Tower height is directly related to the removal efficiency required. Typically, the greater the removal efficiency required, the taller the air stripping tower. However, since structural stability is also a consideration, the tower height can be reduced by increasing the tower diameter. The tower diameter or its cross-sectional area is also a function of the water flowrate and loading rate (flowrate per unit area of tower cross-section). As a rule of thumb, the loading rate typically ranges from 15 to 30 gpm/ft^2.

7.4.7.1 Design Equations

The design equations for an air stripping system are developed based on the following assumptions:

- Steady-state operation
- The influent air is VOC free
- Isothermal operation
- Dilute solution conditions so Henry's law holds true
- Plug flow conditions (no dispersion)

The design equations can be expressed as:

$$Z = HTU \times NTU \tag{7.13}$$

$$\mathrm{HTU} = \frac{L}{K_a} \tag{7.14}$$

$$\mathrm{NTU} = \left(\frac{R}{R-1}\right)\ln\left[\frac{(C_{inf}/C_{eff})(R-1)+1}{R}\right] \tag{7.15}$$

$$R = \frac{H \times G}{L \times P} R = \frac{H \times G}{L \times P} \tag{7.16}$$

where
- Z = packing height (ft)
- HTU = height of transfer unit (ft)
- NTU = number of transfer units (unitless)
- L = liquid loading rate (gpm/ft^2)
- Ka = overall mass transfer coefficient (sec^{-1})
- R = stripping factor (unitless)
- C_{inf} = influent concentration (µg/L)
- C_{eff} = effluent concentration (µg/L)
- H = Henry's law constant (atm)
- G = gas loading rate (CFM)
- P = operating pressure (atm)

The key variables in the equations above are Henry's law constant, H, and the overall mass transfer coefficient, K_a. The Henry's law constant will dictate the applicability of air stripping as a remediation option. Table 7.1 tabulates a list of compounds based on their Henry's law constant and the easiness to strip by air.

From Equations 7.13 and 7.14, it can be seen that the mass transfer coefficient is inversely proportional to the tower height and the packing height. The larger the K_a, the shorter the tower and the packing. As a result, packing materials should be selected so that K_a is maximized. The best way to obtain the value for K_a is to conduct a pilot study at the site, since chemical characteristics will vary from source to source and may affect the stripping process. Figure 7.7 depicts a set-up of a pilot study.

When K_a data is not readily available or pilot testing is not economically feasible, the use of mathematical correlations is the next best alternative. There are two mathematical correlations that are commonly used. The first correlation was developed by Sherwood and Holloway (1940) as

$$K_a = D_L m\left(\frac{L}{u_L}\right)^{1-n}\left(\frac{u_L}{r_L D_L}\right)^{0.5} \tag{7.17}$$

$$D_L = B\frac{T}{u_L} \tag{7.18}$$

where m and B are conversion constants that can be found in the original article published by Sherwood and Holloway. The other terms are defined as follows:

K_a = overall mass transfer coefficient (hr^{-1})
D_L = diffusion coefficient of gas of interest in water (ft^2/hr)
L = liquid velocity (lb water/hr ft^2)
u_L = liquid viscosity (lb/ft hr)
r_L = liquid density (lb/ft^3)
T = absolute temperature (K)

The second correlation was developed by Onda et al. (1968) that takes into account the gas phase resistance and the liquid phase resistance. This correlation also allows the computation of the individual gas phase mass transfer coefficient, k_G, and the liquid phase mass transfer coefficient, k_L. The correlations are expressed as

For liquid phase resistance,

$$k_L\left(\frac{r_L}{u_L g}\right)^{1/3} = 0.0051\left(\frac{L}{a_w u_L}\right)^{2/3}\left(\frac{u_L}{r_L D_L}\right)^{-0.5}\left(a_t d_p\right)^{0.4} \tag{7.19a}$$

and

$$\frac{a_w}{a_t} = 1 - \exp\left[-1.45\left(\frac{s_c}{s}\right)^{0.7}\left(\frac{L}{a_t u_L}\right)^{0.1}\left(\frac{L^2 a_t}{p_L^2 g}\right)^{-0.05}\left(\frac{L^2}{r_L s a_t}\right)^{0.2}\right] \tag{7.19b}$$

For gas phase resistance,

$$\frac{k_G}{a_t D_G} = 5.23\left(\frac{G}{a_t u_G}\right)^{0.7}\left(\frac{u_G}{r_G D_G}\right)^{1/3}\left(a_t d_p\right)^{-2.0} \tag{7.20}$$

The overall mass transfer coefficient is given by:

$$\frac{1}{K_a} = \frac{1}{Hk_G} + \frac{1}{k_L} \tag{7.21}$$

Table 7.1 Stripability of Selected Compounds Based on Henry's Constant

Compounds	Henry's Constant*	
Vinyl Chloride	50	Easiest to remove by air stripping
Tetrachloroethylene	1	↓
Carbon Tetrachloride	1	
1,1-Dichloroethylene	0.6	
Trichloroethylene	0.5	
Toluene	0.3	
Benzene	0.2	
1,1,1-Trichloroethane	0.2	
Chloroform	0.1	
Methylene Chloride	0.1	
Heptachlor	0.06	
1,1,2,2-Tetrachloroethane	0.02	
2,4-Dichlorophenol	0.002	
Nitrobenzene	0.001	
Ammonia	0.0006	Hardest to remove by air stripping
Phenol	0.00005	

*mg/L in air per mg/L in water at 25°C.

where each of the parameter is defined as:

k_L = liquid phase mass transfer coefficient (m/sec)
r_L = liquid density (kg/m^3)
u_L = liquid viscosity (kg/m sec)
g = gravity force (m/sec^2)
L = liquid velocity (kg/m^2 sec)
a_w = wetted packing area (m^2/m^3)
D_L = liquid diffusivity (m^2/sec)
a_t = total packing area (m^2/m^3)
d_p = nominal packing diameter (m)
s_c = critical surface tension of packing material (N/m)
s = liquid surface tension (N/m, s = 0.073 N/m at 20°C)
k_G = gas phase mass transfer coefficient (m/sec)
D_G = gas diffusivity (m^2/sec)
G = gas velocity (kg/m^2 sec)
u_G = gas viscosity (kg/m sec)
r_G = gas density (kg/m^3)
K_a = overall mass transfer coefficient (sec^{-1})
H = Henry's law constant (unitless)

Still, if K_a cannot be calculated with either one of the correlations because of the unavailability of certain parameters, a K_a value of 0.5 to 1.0 per hour can be used as a general rule of thumb for K_a estimation.

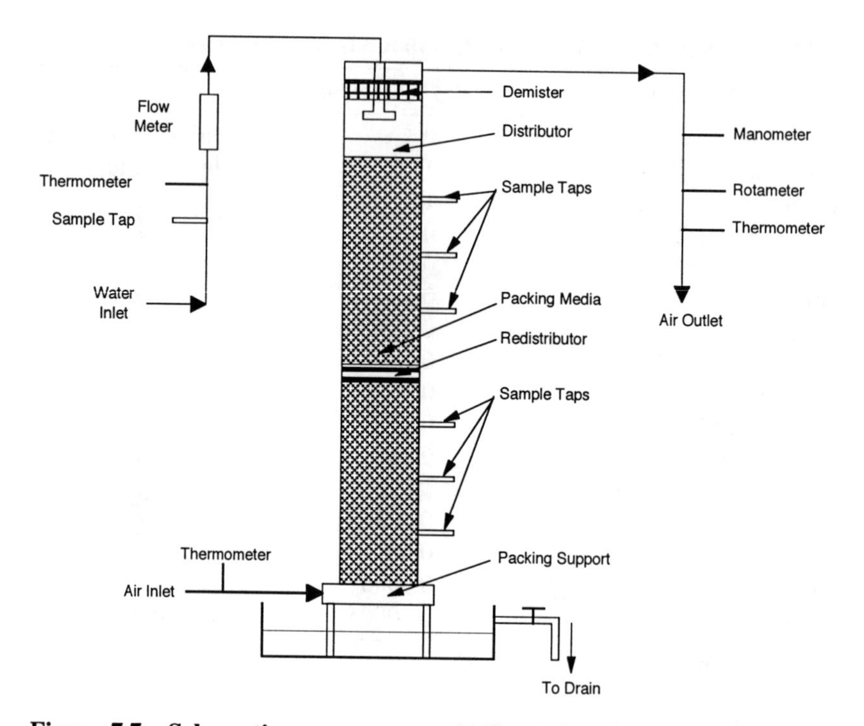

Figure 7.7. Schematic arrangement of pilot scale system.

7.4.7.2 Column Components

A typical air stripping tower consists of a tower shell, demister, water distribution system, packing media, packing support plate, tower sump, and air delivery system as shown in Figure 7.8. Fiberglass is the most common material used to construct short- to medium-height towers because of its good corrosion resistance under most chemical environments. However, fiberglass is not suitable for tall air stripping towers because it is relatively brittle, thus making it somewhat susceptible in strong wind situations. As a result, for tall air stripping towers, other materials such as stainless steel, coated carbon steel, or PVC are generally used. All of these materials offer good corrosion resistance, but their cost is several times higher than fiberglass.

A mist eliminator is placed on top of the tower to prevent the discharge of water droplets. Below the demister is the water distribution system. Water is pumped into the water distributor where it is distributed across the surface of the packing. The most commonly used water distributor is a spray nozzle.

The next item in the tower is the packing media. This is the most important component of an air stripping tower. The main criterion for packing materials is to maximize the surface area for air and water to interact. The packing should also have a large void area to minimize the pressure drop across the tower. There are many types of packing media that are commercially available.

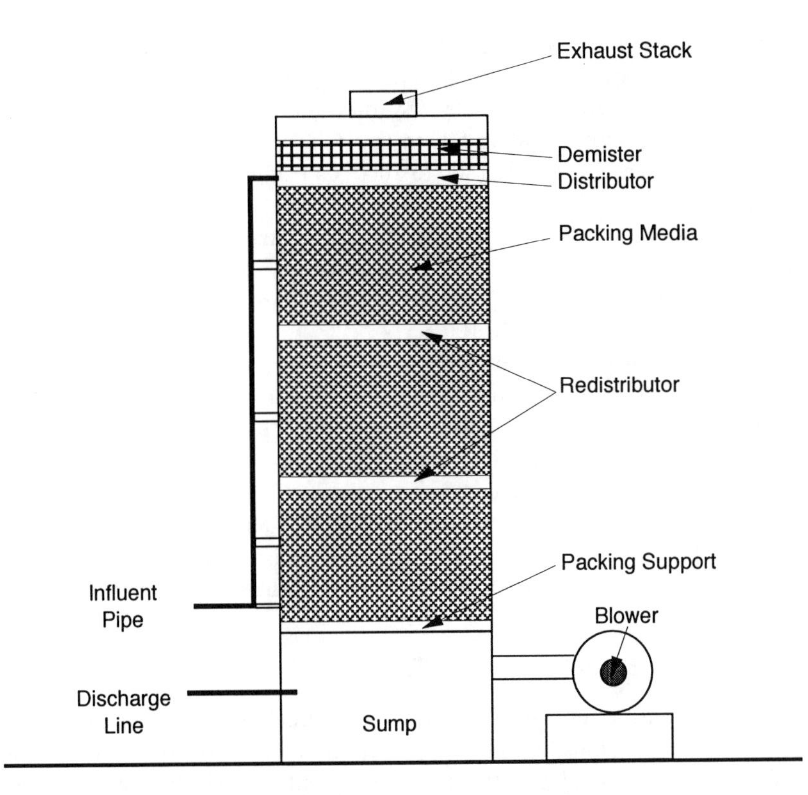

Figure 7.8. Typical air stripping tower components.

The most commonly used packing materials are plastic packings. Plastic packings are chemically inert and will not degrade when exposed to most chemicals encountered in groundwater contamination.

The packing media is supported by a packing support plate. The packing support plate is a metal plate with many openings to allow water flow. The openings of the packing support plate should be just small enough to prevent the packing media from falling into the tower sump.

The bottom of the tower is the tower sump or base. Water from the base typically flows by gravity to discharge in a stream or sewer. The blower makes up the last component of the air stripping tower system. The blower selected should be able to deliver a range of air flow to allow for flexibility in tower operation.

7.4.8 Activated Carbon Unit Design

7.4.8.1 Activated Carbon

Any organic material with a high carbon content (wood, coal, coconut shells, etc.) can be used as the raw material for making activated carbon. The specific process for producing activated carbon varies depending on the raw material used; however, in simple terms, it generally involves grinding the raw material to a suitable particle size followed by carbonizing in a furnace at temperatures approximating 500 to 600°C. The raw material is then activated at a higher temperature (800 to 1000°C) using either a steam or chemical process.

The finished product has an incredibly large internal pore volume per unit particle volume and a network of submicroscopic pores in which physical adsorption takes place. One pound of activated carbon contains an effective surface area of approximately 100 acres.

In the physical adsorption process, molecules are attracted to the carbon surface by weak forces known as van der Waals forces. Generally the surface of the carbon prefers large, bulky, organic molecules with no charge. The molecules must be brought into contact with the carbon surface through a suitable medium such as water.

Many factors affect the adsorption process, such as water quality, temperature, flowrate, concentration of the target adsorbate(s), type of adsorbate(s), number of adsorbates, and level of adsorbate(s) removal required.

7.4.8.2 Mass Transfer Zone (MTZ)

Carbon adsorption systems used in water treatment typically use downflow fixed bed columns whereby the process flow enters the top of the column and flows down through the stationary carbon bed and exits through a strainer at the base of the vessel.

To understand the dynamics of the adsorption process within the fixed bed one must accept the concept of the mass transfer zone (MTZ). The MTZ is the area within the carbon bed where adsorption takes place. The MTZ delineates the regions of the carbon bed where the carbon has reached its adsorption capacity, and where it has yet to begin adsorption. The MTZ moves through the bed in the direction of flow. The depth and rate of movement is dependent on the flowrate and adsorptive capacity (x/m). Breakthrough occurs when the MTZ reaches the effluent zone of the bed (Figure 7.9). If the process flow contains multiple adsorbates, each may have its own MTZ.

Predictable migration of the MTZ requires stable and continuous influent process flows. If the process flow stops for a significant period of time, the MTZ may equalize throughout the bed. Given sufficient down-time, desorption may occur near the more saturated influent zone of the bed and release target adsorbates upon resumed operation. Likewise, if the flowrate increases significantly, the MTZ can lengthen and exceed the depth of the carbon bed, thus providing insufficient residence time for complete removal of the target adsorbate(s).

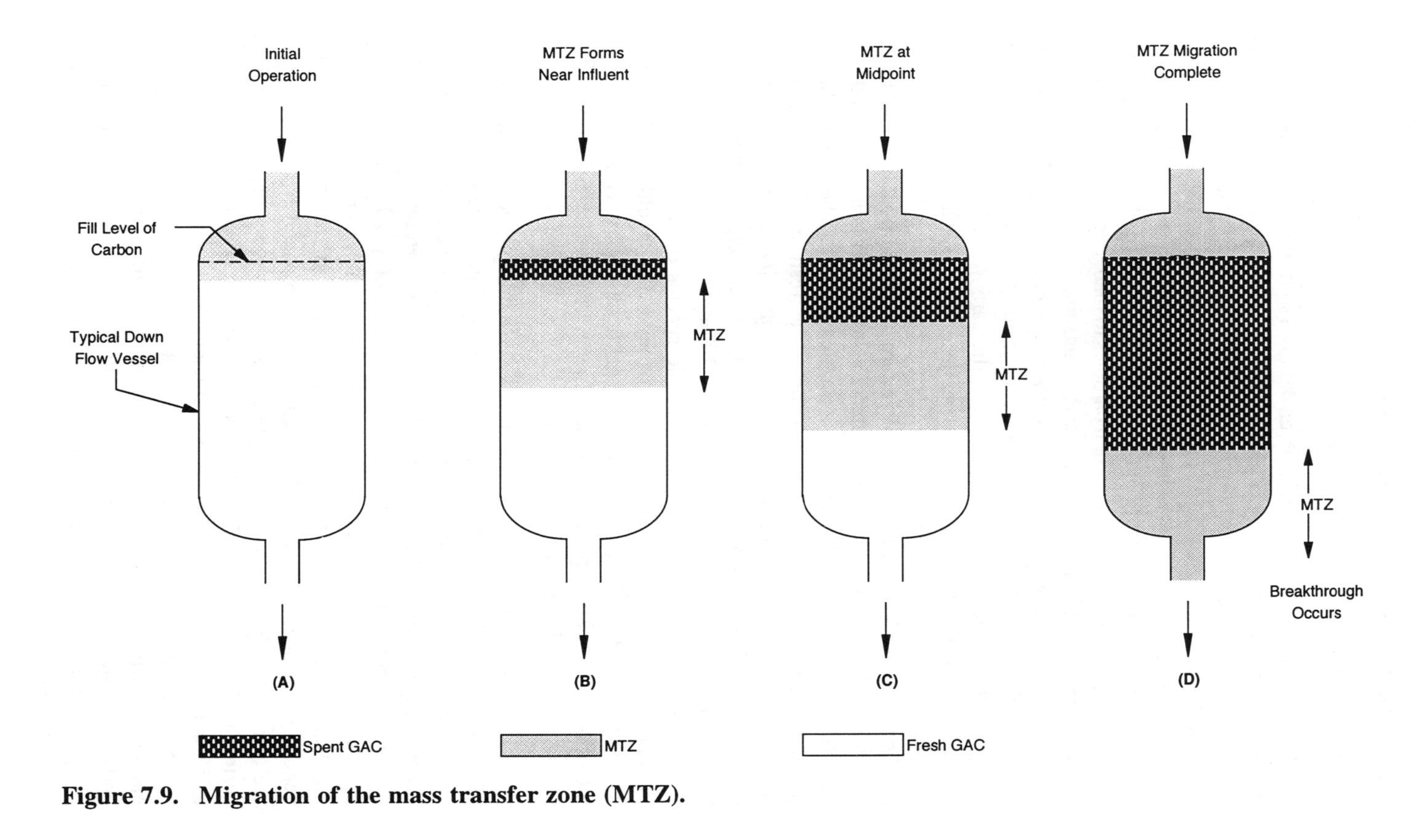

Figure 7.9. Migration of the mass transfer zone (MTZ).

7.4.8.3 Factors Affecting Performance

Water Quality

Total Dissolved Solids. Dissolved solids (solubilized inorganic compounds) are not readily adsorbed during the adsorption process and typically do not present a problem; however, the presence of inorganic salts in the process flow can lead to precipitation of insoluble compounds on the surface of the carbon particle. This accumulation of precipitate can lead to pore blockage and result in reduced carbon adsorption capacity. The presence of insoluble precipitates (scale) may also cause the carbon particles to bind together making carbon "change-out" difficult and render the carbon unsuitable for reactivation.

During initial operation of fresh carbon, the pH of the effluent process flow may rise above the normal range of the influent. This increase in the pH can promote precipitation of scale within the carbon vessel and equipment that follows such as discharge piping or reinjection wells. This occurs because chemical species like sulfates, nitrates, and chlorine adsorb more readily than alkaline species. The effluent pH will return to the normal influent range depending on water characteristics after 250 to 350 carbon vessel volumes have been processed. If this presents a problem, acid can be injected in the influent process flow to maintain effluent pH or the fresh carbon can be initially hydrated using a sodium sulfate solution (the quantity of sodium sulfate in solution should equal approximately 2% of the total weight of the carbon to be treated) and allowed to stand for 10 to 12 hours.

To estimate the effects of the inorganic species on the carbon prior to initial operation, the Langelier Index (scaling potential) should be determined. The index value can be determined if the pH, temperature, calcium content, total dissolved solids, and alkalinity are known. A positive index value may indicate a scaling potential.

Total Suspended Solids. Process flows containing even small quantities of suspended solids can lead to particulate loading. Particulate loading is when suspended solids (>10 microns) are present in the influent process flow and become impinged upon the carbon within the lead adsorber resulting in a higher than normal pressure drop across the adsorber. Generally only the first 6 in. of carbon is affected, thus, backwashing can usually restore adsorber performance.

Backwashing should not be relied upon as the primary method for prevention of particulate loading. Over backwashing an adsorber vessel can mix the carbon and disrupt the MTZ migration. A properly designed prefiltration system is the preferred method. On a smaller system (<10 gpm) a single 10- to 50-micron particulate filter may be sufficient.

Suspended solids finer than 10 microns can create problems on the effluent end of the carbon adsorber as well. These smaller solids can pass through the carbon bed and may contain dissolved or attached target adsorbates. Effluent samples containing these small solids can give the appearance of premature breakthrough.

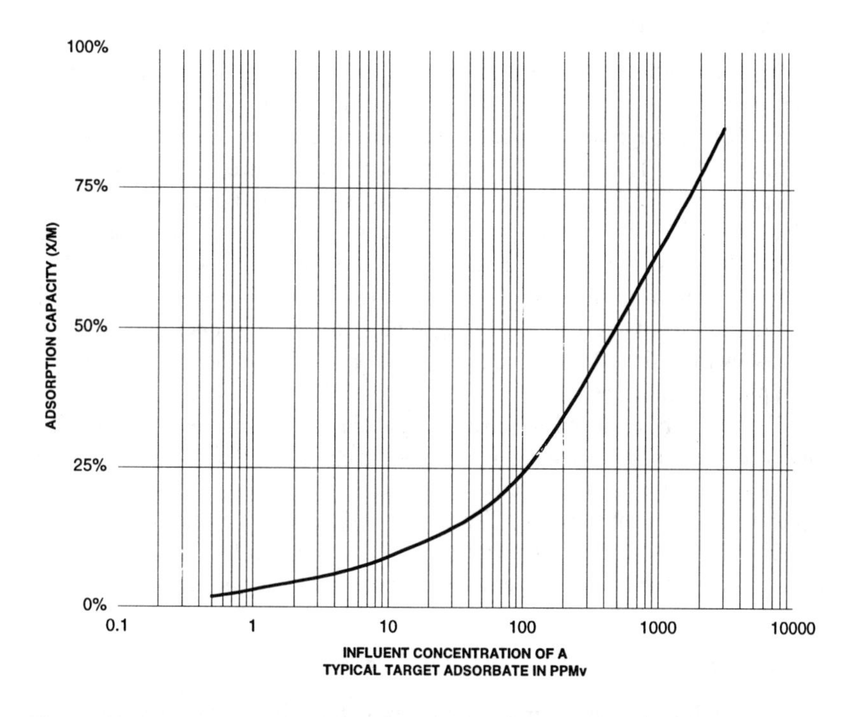

Figure 7.10. Adsorption capacity versus influent concentration.

Concentration of Target Contaminant. The influent concentration for a given target contaminant (adsorbate) will result in a specific carbon adsorption capacity. The greater the influent concentration of the adsorbate, the greater the adsorptive capacity (*x/m*) of the carbon. This occurs because adsorption of a given adsorbate will occur to the point of equilibrium in solution. If, for example, a given adsorbate is present in the process flow at 1 ppm_v, the carbon may yield an adsorptive capacity of 3%; however, if the same adsorbate was present in the process flow at 10 ppm_v, the carbon may yield an adsorptive capacity of 9% (Figure 7.10). Most carbon vendors (as a marketing tool) will provide the adsorptive capacity for specific compounds at specific flow parameters upon request.

Due to the equilibrium effect, desorption can occur if the influent adsorbate concentrations present in the process flow decrease or fluctuate significantly. However, desorption occurs much slower than adsorption and should only be accounted for if the predicted flux is greater than 25% of the design concentration.

The accepted method for safe guarding effluent quality is to design a system based on multiple adsorbers operated in series and perform routine chemical monitoring at influent, intermediate (between adsorbers), and effluent points. The frequency of the monitoring should be greater than the predicted time for lead vessel breakthrough.

Short Circuiting

Short circuiting (channeling) is when the process flow selects a preferential path through the carbon bed. This hinders uniform dispersion of the process flow throughout the area of the bed and disrupts the uniform migration of the MTZ. The problem typically occurs when a fresh carbon adsorber is placed in down-flow service without properly deaerating the carbon. A typical bed of fresh carbon will have pore volume of 40% of the adsorber vessel volume. If the carbon is dry, the pore volume will contain air. This air must be removed prior to placing the adsorber in operation. The air will not migrate out of the carbon during typical down-flow operation because the downward fluid velocities usually equal or overcome the buoyancy of the air trapped in the pores.

The air can be removed and channeling avoided by filling the adsorber vessel with clean water prior to placing the adsorber in operation. Generally a minimum wetting period of 24 hours is required at normal temperatures, although 72 hours is preferred for complete wetting. To further ensure that all air has been purged, the adsorber vessel can be backwashed with 2 to 3 bed volumes of clean water in an up-flow direction at a superficial velocity of 2 to 3 gpm/ft^2.

Air Binding

Air binding is when air accumulates within the adsorber vessel. The accumulation of air begins to displace the process fluids in the top of the adsorber. If allowed to continue the top of the carbon bed can become exposed and begin to dewater. If this occurs the downward migration of the MTZ through the carbon bed will be accelerated thus contributing to premature breakthrough.

A well thought-out piping and extraction system design can manage most air binding problems. The trapped air usually finds its way into the vessel via the extraction system as the result of pump maintenance, sample collection or faulty level control equipment. Well placed air purging valves throughout the system can remove the air prior to its entrance to the vessel. The vessel can be equipped with an air purging valve as well. This is discussed in greater detail later in this section.

7.4.8.4 Carbon Adsorber Process Design

Adsorber Selection

Carbon adsorbers used for groundwater remediation are typically of a down-flow, fixed bed, reactor design. They are commercially available, classified by flow capacity, pressure rating, and weight of dry carbon within the vessel. A few of the most common sizes are 200, 900, 2,000, 10,000, and 20,000 pound units. The larger units (10,000 lbs and above) can be provided in pairs preassembled on a structural skid with all necessary piping and controls.

The criterion for selection typically focuses on desired flowrate and influent concentrations of target adsorbates, which together can provide the estimated daily carbon usage in pounds/day. The goal is to specify a vessel configuration with an acceptable capital cost and maximum time between carbon change out.

Calculated Adsorptive Capacity. The first step in adsorber selection is to determine the site-specific adsorption capacity. This is best measured with on-site pilot testing; however, it may be calculated using the empirically derived Freundlich isotherm which is defined as follows:

$$x/m = K(C_{in})^{1/n} \tag{7.22}$$

where x/m = amount of adsorbate adsorbed per unit weight of carbon
K and n = empirical constants
C_{in} = influent concentration of target adsorbate

The empirical constants K and n can be obtained for most toxic organics from carbon retailers or the EPA 1980 publication *Carbon Adsorption Isotherms for Toxic Organics* (EPA-600/8-80-023).

The calculated method should be used cautiously since it cannot accurately account for the synergetic effect of multiple adsorbates, which may be present in the process flow nor does it consider the effects of dissolved and suspended solids. It does, however, have value as a pre-design screening tool to determine if carbon adsorption is an effective treatment option.

Measured Adsorptive Capacity and Pilot Testing. Short-term pilot tests provide very accurate values for *x/m*. Generally a test is conducted at some fraction of the design flow and would employ two small adsorbers in series with a minimum of process controls (Figure 7.11). Duration of the test is typically the period of time required to load the lead vessel to the point of breakthrough. Monitoring during the test may include the collection of groundwater quality data as well as groundwater elevations (often a pump test may coincide with the pilot test). Following the test a sample of the saturated carbon is retained for laboratory analysis by method ASTM D 1375. This analysis will provide the measured *x/m* achieved by the carbon in the pilot adsorbers. For this reason it is important that the carbon used for the pilot test be of the same grain size and quality as that which will be used for the full scale adsorbers.

Once the adsorptive capacity (*x/m*) is obtained, the carbon usage in pounds/day at the full scale flowrate can be determined using the following equation. The equation is presented in expanded form so units can be tailored to individual needs.

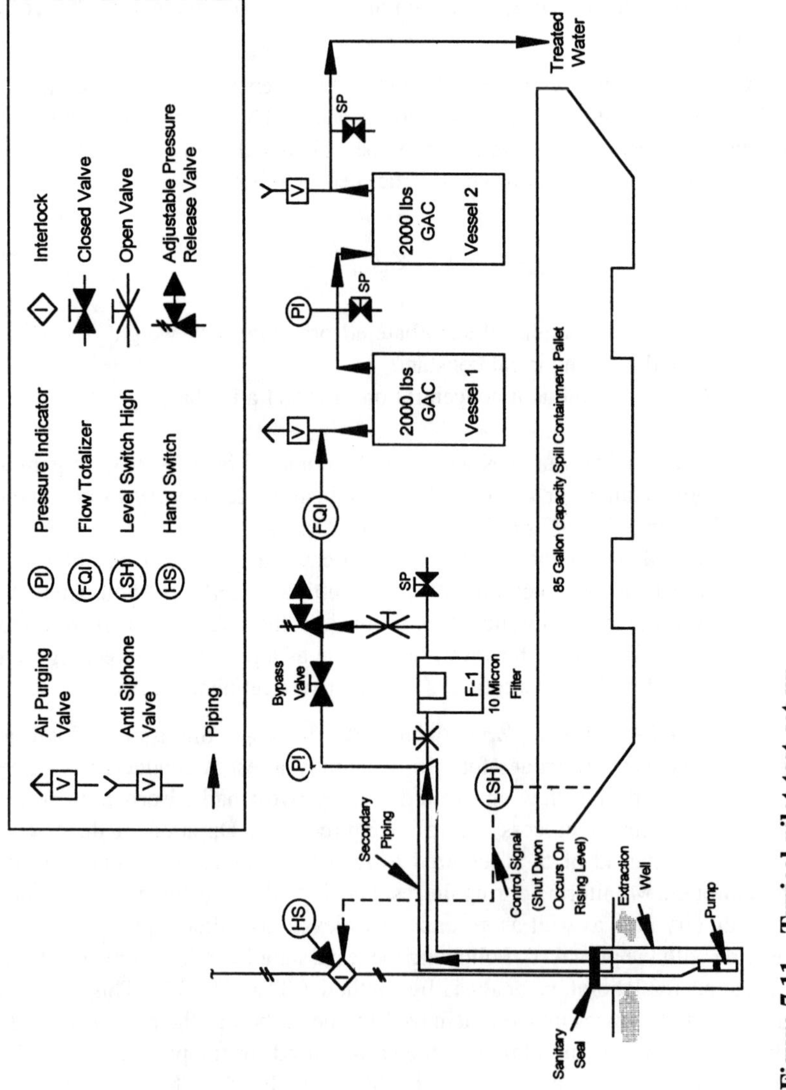

Figure 7.11 Typical pilot test set up.

$$C_d = (C_{in} - C_o)(Q)\left(\frac{1.0\text{ gm}}{1,000\text{ mg}}\right)\left(\frac{1,000\text{ L}}{264\text{ Gal}}\right)\left(\frac{0.0022\text{ lb}}{1.0\text{ gm}}\right) \quad (7.23)$$

$$G_d = C_d/(x/m)_b \quad (7.24)$$

where C_d = mass of target adsorbate/s in pounds/day
C_{in} = influent concentration of target adsorbate in mg/L
C_o = effluent concentration of target adsorbate in mg/L
Q = process flow rate in gallons per day
G_d = carbon usage in pounds/day
$(x/m)_b$ = measured amount of adsorbate per unit weight of carbon in mg/gm at breakthrough

Once the carbon usage in pounds per day is determined adsorber selection is a simple matter of plotting the capital and operating cost of the vessel(s) versus time. Large adsorbers may have a higher capital cost; however, the frequency of carbon exchange and required monitoring will be relatively low. Additionally, the carbon vendor may offer a lower price for delivery and sale of large volumes of carbon (Figure 7.12).

Superficial Velocity. Following the initial adsorber selection the superficial velocity of the process flow within the vessel should be determined to confirm the selected adsorber size and configuration is adequate for the full scale design flow. To begin, the physical specifications for the vessel must be obtained from the manufacturer. The specifications may include velocities at specific influent flows for a given carbon density and grain size. If not, the velocity can be approximated using the following equation.

$$V_v = Q_{in}/A_v \quad (7.25)$$

where V_v = the superficial velocity within the carbon vessel in gpm/ft^2
Q_{in} = full scale influent flow rate in gpm
A_v = the cross-sectional area of the adsorber in ft^2 minus approximately 60% for area occupied by carbon

In general, velocities at or below 8 gpm/ft^2 are acceptable. Velocities greater than 9 gpm/ft^2 should be avoided unless approved by the vessel manufacturer. High superficial velocities can lead to bed compression, excessive head loss and may not provide sufficient residence time for complete removal of the target adsorbates.

Selection of the adsorber(s) is merely the first in a series of decisions in the design process. The remediation design engineer must account for site variables and long-term maintenance of the system.

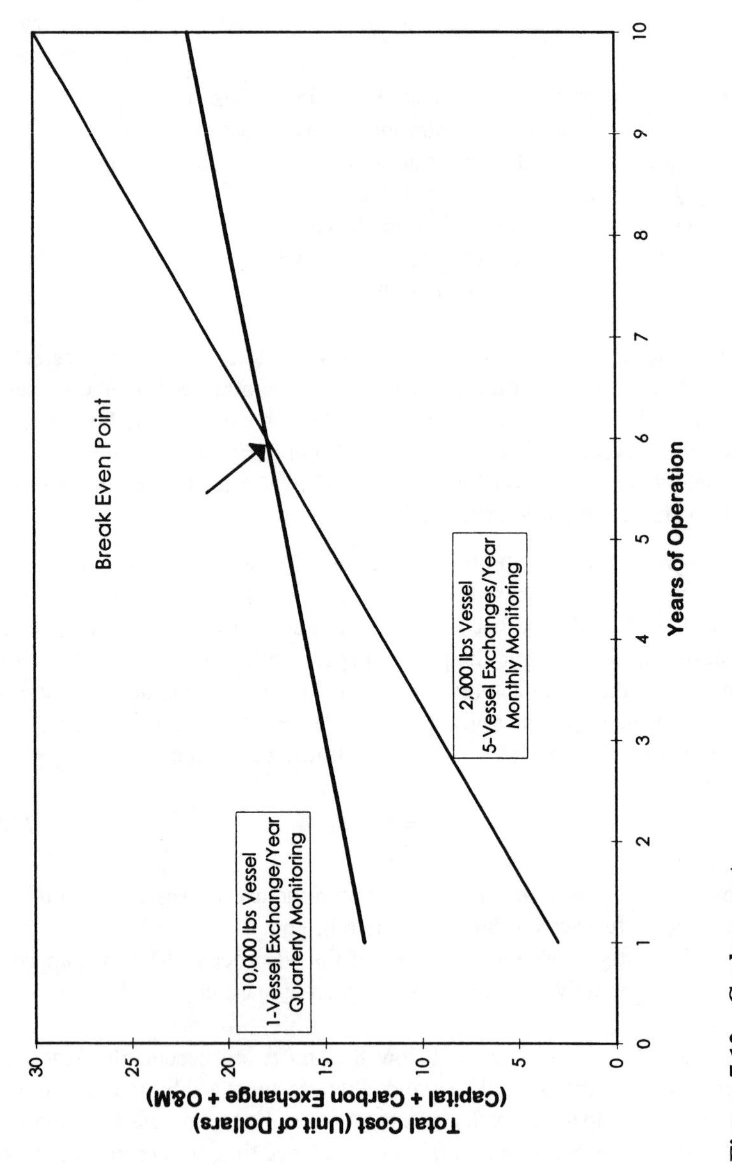

Figure 7.12. Carbon cost.

Piping. In most applications the final adsorber selection will include a pair of same size vessels configured in series to ensure effluent quality and provide the ability to exhaust all carbon in the lead adsorber. Once the lead adsorber is exhausted the lag adsorber should take over the lead position and the lead adsorber now filled with fresh carbon should become the lag adsorber. If the adsorbers were not rotated a low concentration MTZ could develop and begin a slow migration within the lag adsorber.

Figures 7.13 through 7.16 depict a basic process piping design that provides the ability to alter the position of the absorbers in terms of process flow, switch to parallel operation, and take either adsorber off-line without requiring a system shut down. As mentioned previously, the larger systems are typically sold complete with all piping and valves (similar to what was just described).

7.5 DESIGN EXAMPLE

7.5.1 Example 1

The following example assumes all previous assessments, site characterizations, and hydrogeological investigations have already been performed. The design engineer's task is to analyze the available data and design a remediation system that will satisfy all technical and regulatory requirements.

7.5.1.1 Background

The site was the former location of a manufacturing facility which operated during the early to mid-1900s. Residuals of the manufacturing process were detected in the soil and groundwater at the site at levels sufficient to warrant remedial action.

A detailed study followed that outlined two remedial objectives: (1) remove on-site impacted soil and (2) contain off-site migration of groundwater containing listed constituents. As a result, affected soil was excavated and taken off-site and a detailed study of the groundwater behavior completed.

7.5.1.2 Previous Findings and Conclusions

As a result of the extensive groundwater study conducted at the site, the following site information and design objectives were established:

- Groundwater gradient curves to the northeast as a result of off-site pumps operated by others.
- Groundwater elevation fluctuates 1 to 2 feet daily.
- The delineated plume narrows as the gradient slope increases near the northeast corner of the site.
- Influent concentrations of total organics fluctuates between 1 and 30 mg/L (mostly total petroleum hydrocarbons and polycyclic aromatic hydrocarbons).

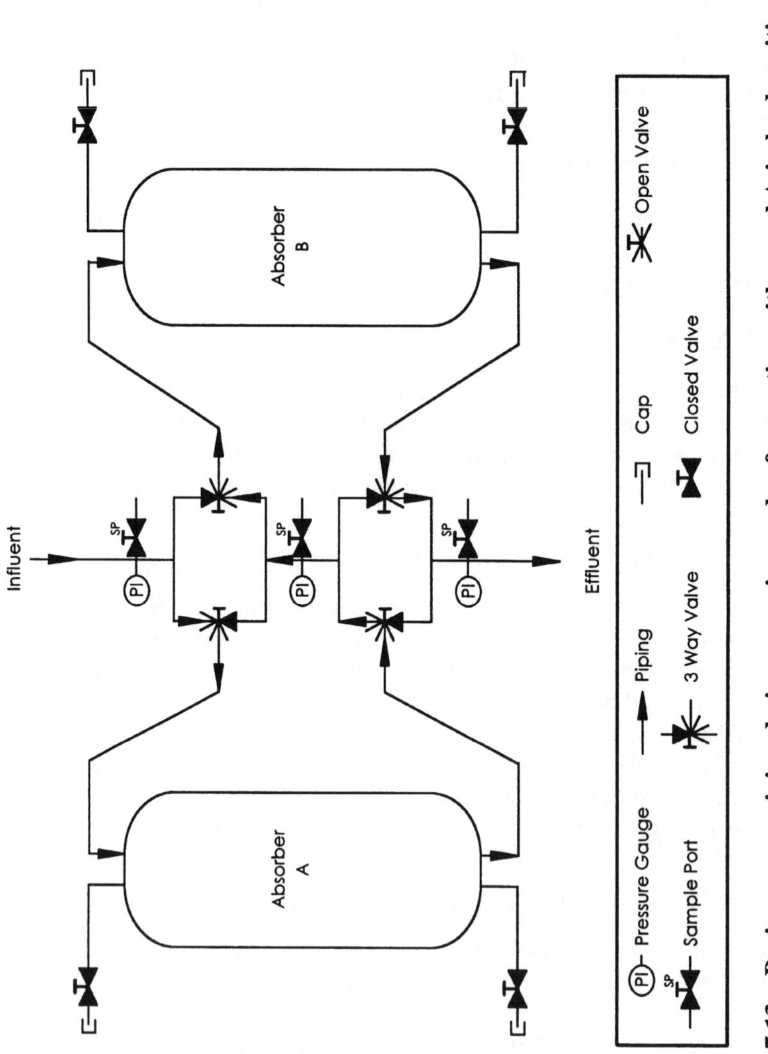

Figure 7.13. Basic process piping design — series mode of operation with vessel A in lead position.

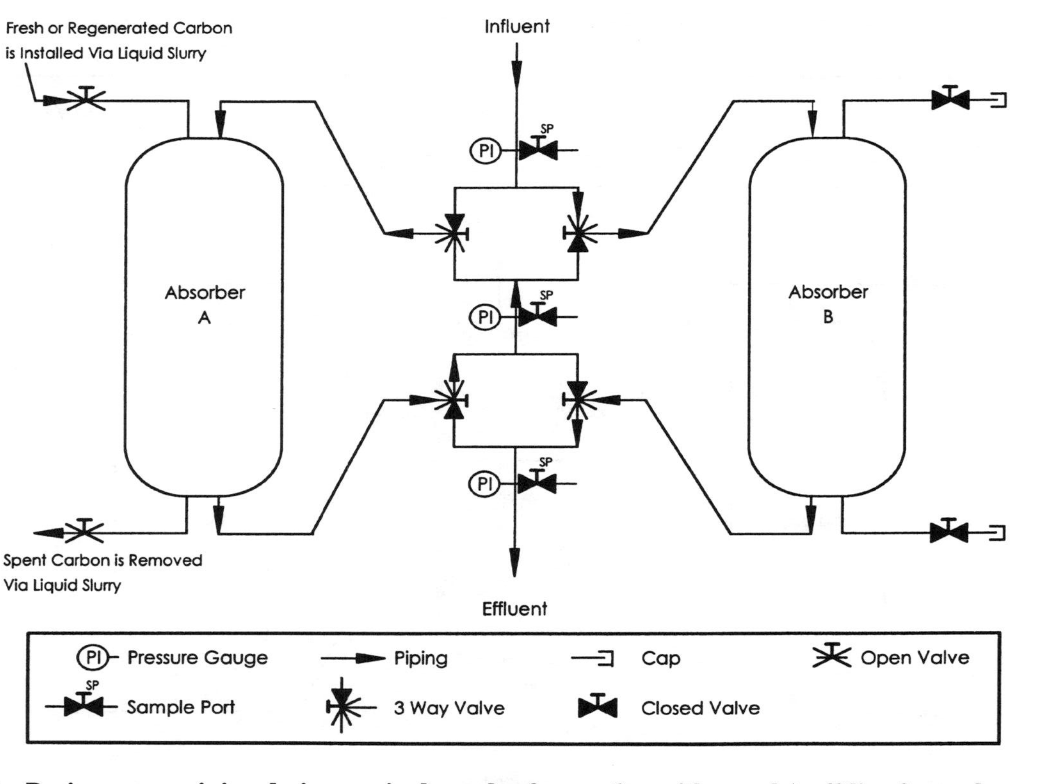

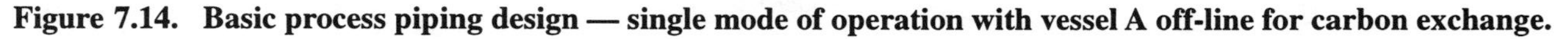
Figure 7.14. Basic process piping design — single mode of operation with vessel A off-line for carbon exchange.

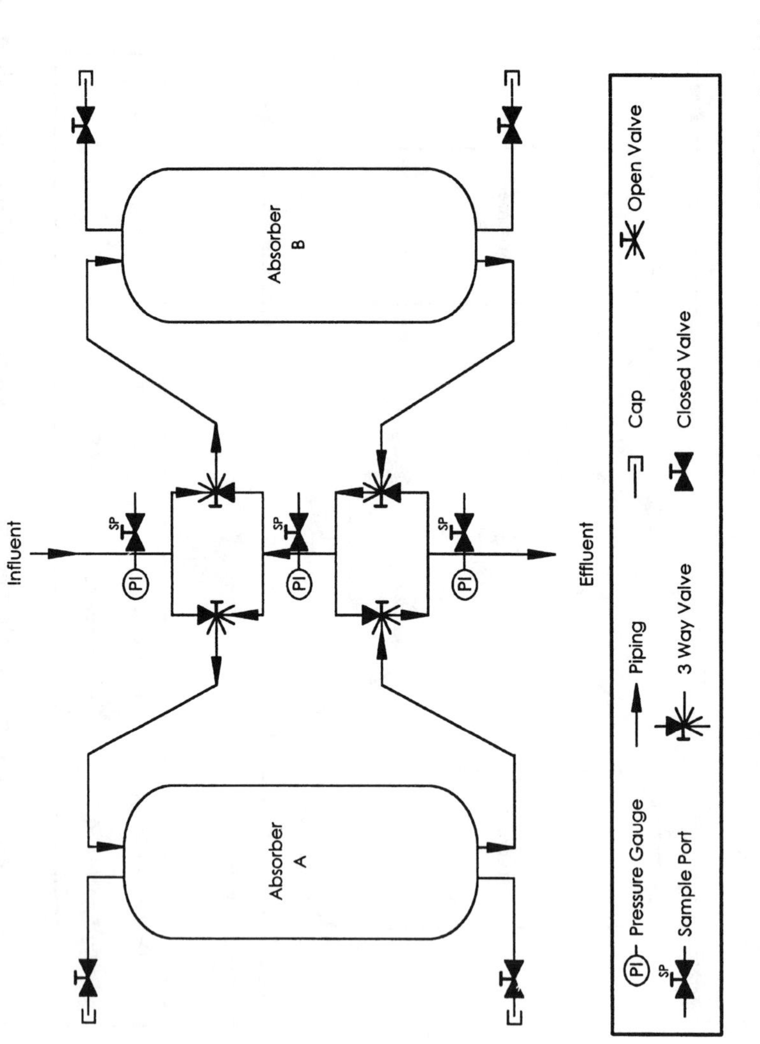

Figure 7.15. Basic process piping design — series mode of operation with vessel B in lead position.

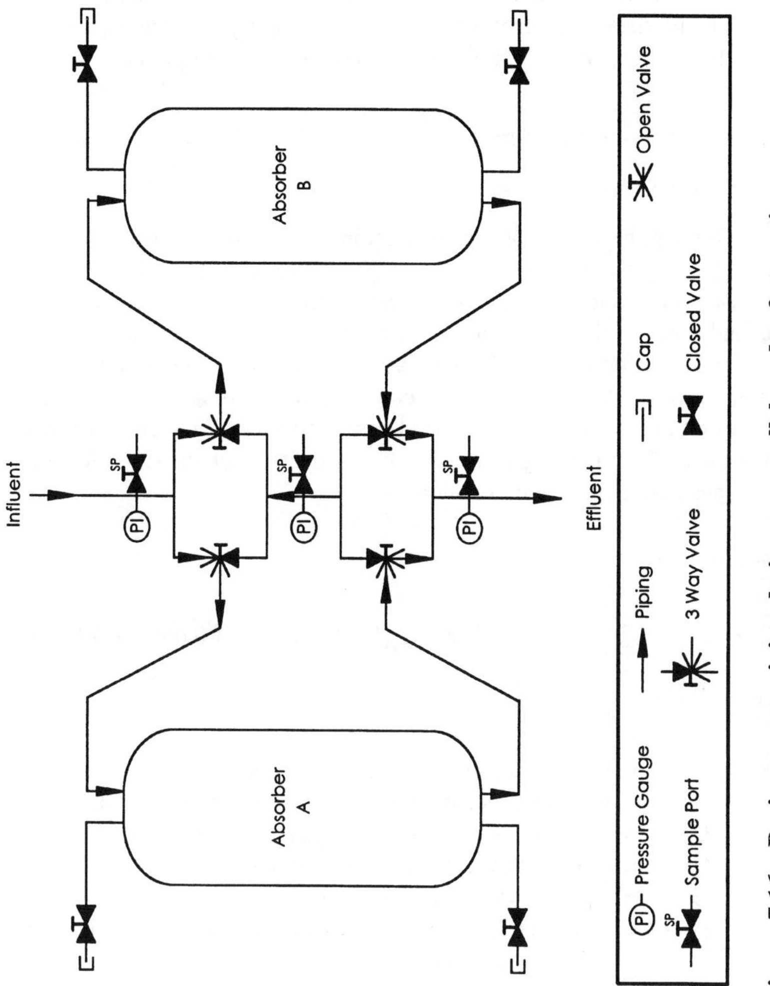

Figure 7.16. Basic process piping design — parallel mode of operation.

- A 3-week carbon adsorption pilot test determined the measured $(x/m)_b$ to be 20 g/100 g.
- Total dissolved and total suspended solids were each low, therefore, scaling within the adsorbers will not be a problem and prefiltration requirements will be standard.
- Pump test data collected during the pilot test, together with other data, indicated that an array of 4, possibly 6 extraction wells along the northeast property boundary, each pumping continuously at 10 to 15 gpm, should control off-site migration of the plume.

7.5.1.3 Conceptual Design

Carbon Adsorber Selection

Following the review of previous findings and conclusions, a conceptual design for a groundwater extraction and treatment system (GWETS) was completed and submitted. The design was based on a 30-year operating life-span and satisfied the requirements outlined in the preliminary design.

The design for the GWETS was based on the adsorber system, which was a preassembled, skid-mounted pair, complete with piping for backwashing, carbon exchange, and alternating lead vessel position. Each adsorber vessel contained 10,000 lb of virgin coconut shell-activated carbon. The calculations for the adsorber system are as follows:

Equation:

$$C_d = (C_{in} - C_o)(Q)(1.0\ \text{gm}/1000\ \text{mg})(1000\ \text{L}/264\ \text{gal})(0.0022\ \text{lb}/1.0\ \text{gm})$$

$$G_d = C_d/(x/m)_b$$

where C_d = mass of target adsorbate/s in pounds/day
C_{in} = influent concentration of target adsorbate in mg/L
C_o = effluent concentration of target adsorbate in mg/L
Q = process flow rate in gallons per day
G_d = carbon usage in pounds/day
$(x/m)_b$ = measured amount of adsorbate per unit weight of carbon in mg/gm at breakthrough

Calculation:

$$C_d = (25 - 0.05)(57,600)(1.0\ \text{gm}/1000\ \text{mg})(1000\ \text{L}/264\ \text{gal})(0.0022\ \text{lb}/1.0\ \text{gm})$$

$$= 11.97$$

$$G_d = 11.97/0.2 = 59.88$$

The daily carbon usage in lb/day is roughly 60. The expected lead adsorber exchange frequency based on the carbon usage is 167 days.

The superficial velocity of the process flow within the vessel was determined as follows:

Equation:

$$V_v = Q_{in}/A_v$$

where V_v = the superficial velocity within the carbon vessel in gpm/ft²
Q_{in} = full scale influent flow rate in gpm
A_v = the cross-sectional area of the adsorber in ft² minus approximately 60% for area occupied by carbon

Calculation:

$$V_v = 40/20.1 = 1.99$$

The calculated superficial velocity of the process flow within the vessel is roughly 2 gpm/ft².

Well Location

The location for the extraction wells was selected based on previously collected data and for the most part followed the gradient contour in the area of the downgradient edge of the plume. The reinjection wells were not originally modeled as part of the remedial effort but were added to the design later as a discharge alternative to the sanitary sewer. They were located on-site as far upgradient as possible (Figure 7.17).

Piping and Controls

Since the adsorber system would arrive with its own ready-to-go piping and valves, the remaining process design focused on the extraction and discharge manifolds and process instrumentation location.

The extraction wells would each be equipped with a submersible pump, which would receive its power via a variable frequency motor drive (VFD). The VFD would allow electronic control of the pumps rotating speed by digitally recreating a variable alternating current wave form, thus allowing precise and independent electronic flow control at each extraction well (Figures 7.18 and 7.19).

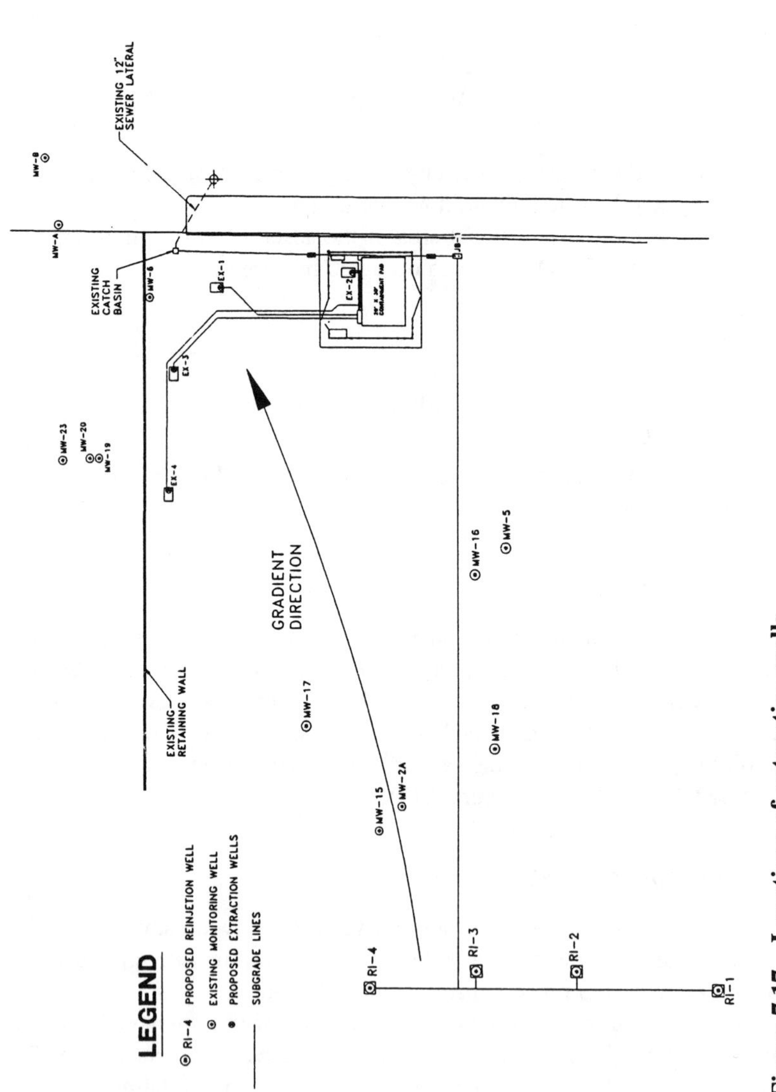

Figure 7.17. Locations of extraction wells.

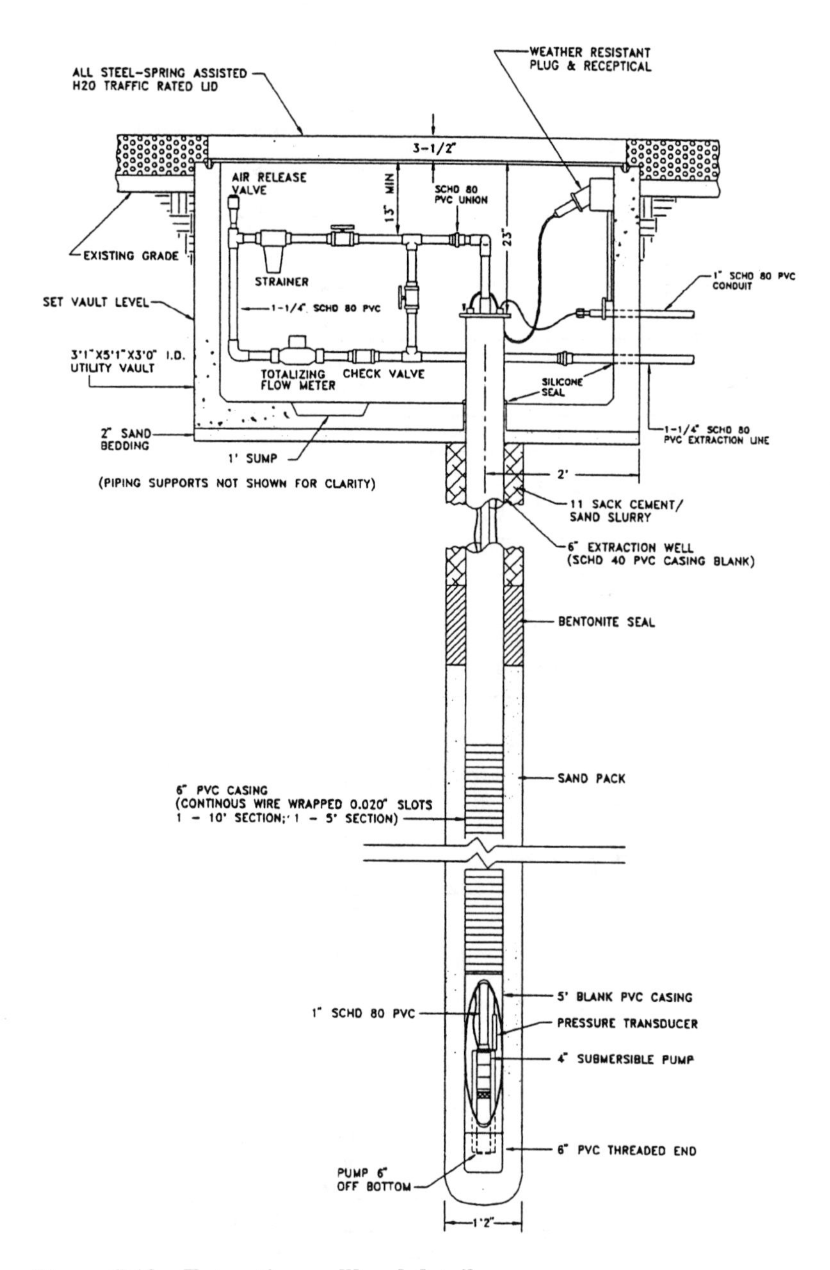

Figure 7.18. Extraction wellhead details.

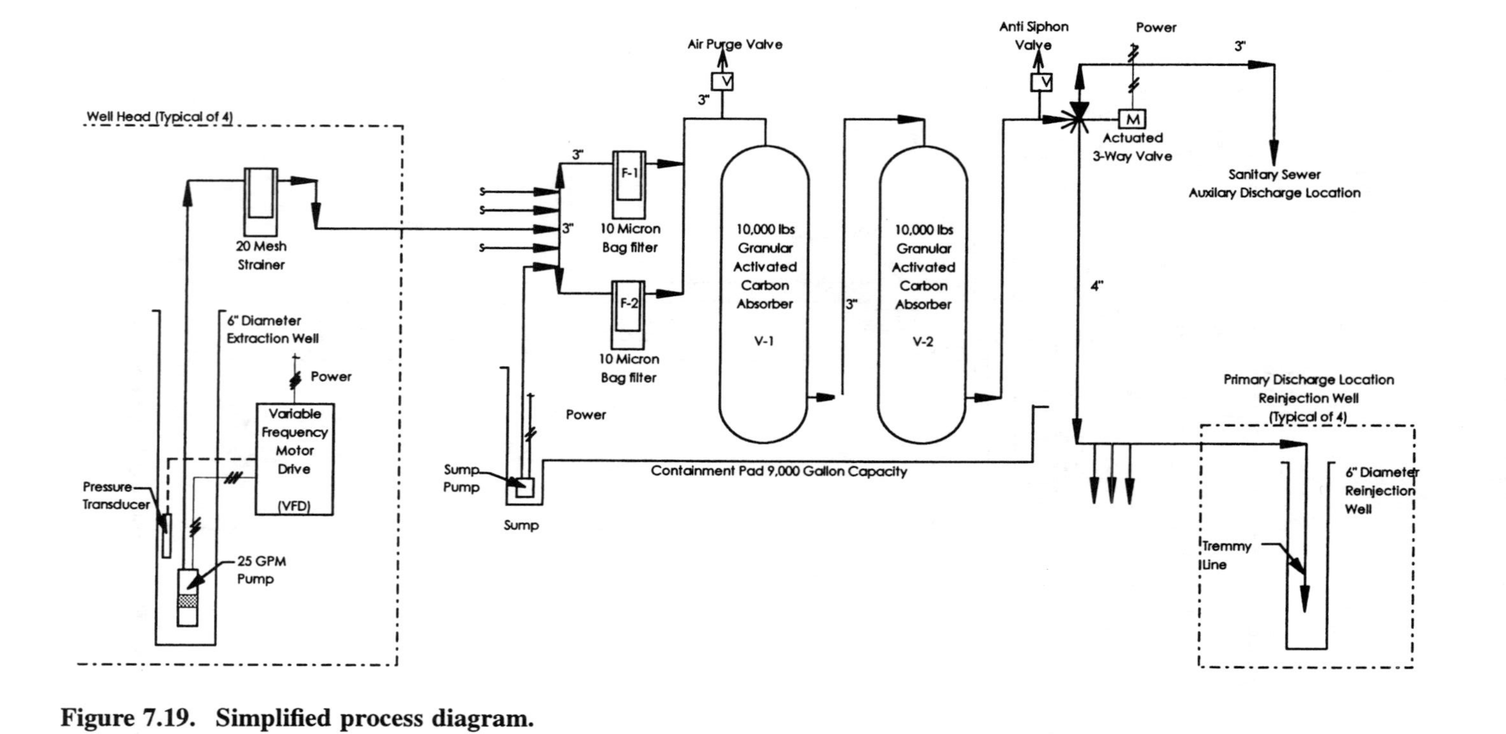

Figure 7.19. Simplified process diagram.

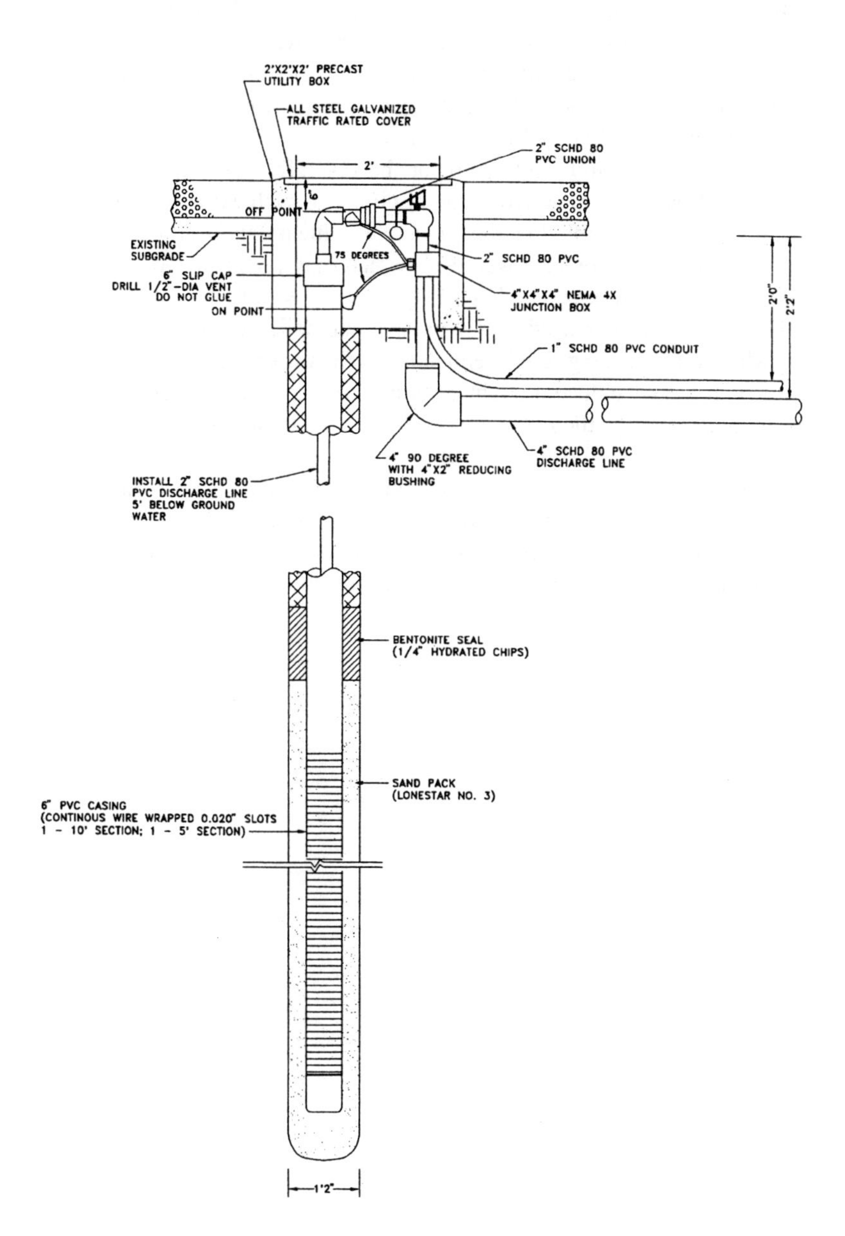

Figure 7.20. Injection wellhead details.

The GWETS would receive a central processing unit (CPU) that would control the VFDs and dependent flow at the extraction wells as well as monitor the system's various process variables (i.e., line pressure groundwater elevations, line velocities, etc.) as well as the status of off-site pumping activity performed by others. The CPU would also communicate to the engineer and others via telephone line and radio telemetry.

The extraction piping manifold for the GWETS was located within the containment pad. This required separate piping to each extraction well; however, this made it possible to collect discrete influent data from the individual wells without accessing each wellhead, which may not always be convenient or possible due to future vehicle parking.

The discharge manifold was a passive subgrade system designed to be somewhat self-balancing. Each reinjection well was equipped with a tremmy line, which would induce a slight negative pressure on the manifold during reinjection, providing the reinjection well with the least mounding (or highest vacuum) the most flow (Figure 7.20)

If flooding occurred at the reinjection wellhead, a mechanical float-actuated valve would stop flow and an electronic float switch would signal the CPU. If two or more reinjection wells were flooded, the CPU would redirect treated effluent to the sanitary sewer and notify the engineer.

8 SUMMARY OF OTHER REMEDIATION SYSTEMS

8.1 INTRODUCTION

Various other technologies or remediation methods beside those described in Chapters 5, 6, and 7 are available for the treatment of contaminated soils and/or groundwater. The majority of these remediation systems can be considered innovative technologies; however, some, like excavation and disposal and site containment, are conventional methods.

This chapter provides a brief summary of some of the technologies presently available to the design engineer. For convenience, they are divided into in situ and non-in situ methods; however, most of the technologies are applicable to both situations. Many of these remedial methods are available as patented technologies and are provided by highly specialized companies.

8.2 IN SITU METHODS

A wide range of in situ remediation technologies are described here, including solidification, vitrification, containment, soil washing, and phytoremediation. These remedial techniques include physical, chemical, and biological processes and, in many cases, a combination of these processes.

8.2.1 Solidification/Stabilization

Solidification refers to the addition of chemicals to the soil structure to produce an inmobile mass. It is also known by various other terms, such as soil stabilization, encapsulation, or chemical fixation. The basis for the remediation is not destruction of the contaminant, but immobilization. Chemicals incorporated into the contaminated soil include Portland cement and other lime-containing products, such as flyash, groundblast furnace slag, or cement kiln dust.

Solidification can also be accomplished by the use of an organic polymer. This technique has been used more extensively for the treatment of chemically contaminated sites, such as hazardous heavy metals, including lead, mercury, barium, cadmium, chromium, selenium, and silver, than for petroleum-contaminated sites. In addition, solidification/stabilization is not quite applicable for the treatment of the lighter hydrocarbons such as gasoline. Diesel, fuel oil, and other heavy hydrocarbons are amenable to the treatment method. Solidification/stabilization techniques are applicable for both in situ and non-in situ situations.

Besides contaminant type, other important considerations in the evaluation of soil stabilization as a treatment alternative are subsurface conditions, site-specific conditions and logistics, and costs. The method is more suitable for sites with tight soils than sites with highly permeable soils. It is only applicable for the treatment of the vadose zone; as such, depth to groundwater will be a consideration. The deeper the depth to the water table, the better it will be. From a practical standpoint, the method will be feasible at a site where the contaminant extent is small rather than large. In situ soil mixing is accomplished with mechanical mixers, standard backhoe buckets, or drilling or augering techniques.

8.2.2 In Situ Vitrification

In the vitrification process, an electric current is passed through the contaminated zone, melting the soil. Once molten, the mass becomes the primary conductor and heat transfer medium, allowing the process to continue. The molten mass extends downward and laterally as long as power is supplied. As the soil cools, the contaminants and the soil matrix are locked into a fused mass.

The process is applicable to soils impacted by volatiles, semivolatiles, nonvolatile organic compounds, and heavy metals. In situ vitrification generally results in a 20 to 40% reduction in volume for typical soils. As cooling continues after the process has been completed, additional subsidence may occur. The treated mass may take several months to a year or more to cool.

8.2.3 Containment Cutoff Walls

Containment walls or subsurface barriers are designed to contain, capture, or divert groundwater flow or the contaminant plume in the vicinity of a site. Containment walls can be constructed as slurry walls, grout curtains, or sheet pilings. Hydraulic barriers created by extraction and injection wells can also be considered a containment system. In some cases, a containment system may include both containment walls upgradient of the contaminant source and a groundwater extraction system downgradient of the plume. Containment walls are typically designed to "anchor" into the bedrock or aquitard; where it does not reach bedrock, it is often referred to as a "hanging wall."

Slurry walls are the most common subsurface barriers because they provide a relatively inexpensive means of reducing groundwater flow. The key word is "reduce" because no barrier can provide 100% impermeability to water flow. Slurry walls are constructed in a vertical trench excavated under a slurry. The slurry acts essentially as a drilling fluid. The different types of slurry walls are differentiated by the materials used to backfill the trench. Usually, an engineered soil mixture is blended with bentonite slurry and placed in the trench to form a soil-bentonite slurry wall.

Grout curtains can be divided into two categories: particulate and chemical. Particulate grouts are composed of Portland cement and/or clay and water. Chemical grouts consist of a chemical base (such as sodium silicate, acrylate, and urethane), catalyst, and water or solvent. Grout curtains are typically keyed into impervious soil formations to form an effective barrier.

Sheet piling is seldom used as a groundwater barrier; it is used more frequently in construction for temporary dewatering purposes or erosion protection of slopes. Sheet piles can be made of wood, precast concrete, or steel. Wood is an ineffective water barrier, whereas concrete is used primarily where great strength is required. Steel is the most effective in terms of groundwater cutoff and cost.

8.3.4 Soil Flushing

Soil flushing, also known as soil washing or soil leaching, refers to the in situ process in which the zone of contamination is gradually remediated by flushing fresh water or a water-surfactant mixture through the soil particles. The water dissolves and carries the contaminants to the water table where they are extracted to the surface for treatment via a pump and treat system. The process is a function of the contaminants' solubilities and Henry's law constants.

Application of this method to petroleum-contaminated sites is based on the principle that petroleum hydrocarbons, though not very soluble, are not completely insoluble. While the Henry's law constants for petroleum hydrocarbons of concern, such as benzene, toluene, ethylbenzene, and xylene, are fairly high (resulting in a tendency to remain in greater concentrations in air when an air/water system is in equilibrium), the introduction of large quantities of flushing water will change the equilibrium of the vapor/liquid partitioning, and can result in some contaminant vapors being solubilized.

Other factors that affect the success of soil flushing include local soil conditions and surfactants added to the flush water. Flushing works best in soils with high permeability, such as gravel or sand. Soils with high silt and clay content impede movement of the flushing solution through the soils, resulting in less removal. Soils with high organic carbon content and high clay content would tend to have stronger sorption characteristics and thus be less suitable for flushing. The process is enhanced if a surfactant or detergent is added to the flushing solution. Surfactants are either natural or synthetic chemicals that have the ability to promote the wetting, solubilization, or emulsification of various organic chemicals. Cost-effectiveness is improved if biodegradable surfactants are used.

From a process standpoint, soil flushing consists of essentially two processing systems: one for preparation of the flush solution or aqueous chemical solution being injected, and another for the fluids being removed from the recovery wells. Water to be used for the injection fluid is treated to reduce hardness if necessary. The water is then blended with surfactants and introduced into the subsurface via injection wells, infiltration galleries, surface spraying, or a percolation network. Recovery or extraction wells are required for pumping the contaminated elutriate to treatment units. Elutriate is the mixture of water, contaminants, and surfactants that is recovered in the soil flushing process. The principal disadvantage to soil flushing is the generation of large quantities of contaminated elutriate that require treatment.

There are other considerations that may limit the feasibility of soil flushing as a remediation alternative. Site-specific factors, such as the presence of underground structures and utilities, will eliminate the possibility of implementing soil flushing. This will probably apply to a great number of leaking underground storage tank sites. The use of surfactants also merits thorough investigation. The groundwater geochemistry should be assessed for naturally occurring constituents that may cause adverse effects to the process, prior to the addition of any surfactant. Typically, the interactions of the surfactant with the chemical, biological, and physical properties of the unsaturated zone are uncertain, and must be determined at each site.

8.2.5 Phytoremediation

Phytoremediation refers to the use of plant root systems for subsurface in situ remediation. The mechanisms of treatment here are twofold: enhanced in situ bioremediation through increased oxidation rates and bacteria-stimulating carbonaceous material, and uptake of the contaminants by the plant's roots (phytoextraction). The vertical extent of remediation is limited by the depth of the plant's root system. The zone occupied by the root system is known as the rhizosphere. Within this zone, bacterial activity is stimulated by carbonaceous material deposited by the plant's root system. Thus plant mass consists of decaying roots and root hairs, as well as secretions from the root system.

8.3 NON-IN SITU METHODS

Non-in situ remedial techniques described here include excavation and disposal, thermal destruction technologies, asphalt incorporation, and soil washing.

8.3.1 Excavation and Disposal

Soil excavation and disposal often lends itself as the best alternative when a quick-fix solution is desired. Soils impacted with petroleum hydrocarbons are dug up, removed from the site, and disposed of in a landfill or at a special facility that accepts contaminated soils for further treatment. Alternately, the soils can be treated on-site where space is available.

Excavation and disposal generally represents the most aggressive approach to any cleanup strategy. It is, however, seldom a cost-effective option and is appropriate only under certain ideal site-specific conditions. Transportation and disposal cost for contaminated soils is expensive and often renders soil excavation a nonfeasible treatment alternative. Site-specific restrictions also present difficulties. Often it is not possible to remove all the contaminated soils due to the presence of structures and utilities. Additionally, the depth to the hydrocarbon-impacted soils may mandate that shoring requirements be addressed, adding significantly to the cost of the project.

8.3.2 Thermal Treatment Technologies (Incineration)

Thermal technologies are processes or systems designed to break down contaminants through either combustion or pyrolysis by exposure of the waste material to high temperature in a controlled environment. This technology can be used for the treatment of contaminated groundwater, soils, sludges, and air streams.

In thermal treatment technology, the contaminated soil is incinerated resulting in pyrolysis of petroleum hydrocarbons to carbon dioxide and water. The most common incineration technology used is the rotary kiln. Other types of incineration are infrared incineration, liquid injection, hearth, fluidized beds, circulating beds, pyrolysis processes, plasma systems, and various kinds of indirect heating systems. An incineration system includes a number of subsystems including waste pretreatment, waste feed, combustion unit, heat recovery, air pollution control equipment, and residue handling and disposal. Rotary kiln incinerators are slightly inclined refractory-lined, rotating cylinders. Rotary kiln incineration involves the controlled combustion of organic contaminants under net oxidizing conditions. Both contaminated soil and supplementary fuel are injected into the high end of the kiln and passed through the combustion zone as the kiln slowly rotates. Rotation of the combustion chamber creates turbulence and improves the degree of combustion. Retention time in the kiln can vary from several minutes to more than an hour. The contaminated soil is substantially oxidized to gases and ash within the combustion chamber. Ash is removed from the process at the lower end of the kiln. Flue gases are passed through a secondary combustion chamber and then through air pollution control units for particulate and acid gas removal.

8.3.3 Asphalt Incorporation

Asphalt incorporation is a variation of the soil stabilization/solidification technique previously described. Contaminated soils are incorporated into asphalt, inmobilized, and used for roadway paving. The quantity of contaminated soils in the asphalt mixture is typically only a few percent of the asphalt. The contaminants in the soil are remediated through a mixture of processes, including immobilization, volatilization, and thermal destruction. Soils heavily impacted by gasoline are not suitable for asphalt incorporation; this technology is more suitable for soils contaminated with heavier hydrocarbons.

8.3.4 Soil Washing

Soil washing is a variation of the soil flushing process whereby the "washing or flushing" operations occur above-ground or ex situ rather than in situ. By performing the washing above-ground in a reactor, contaminants can be more effectively removed because there is greater overall control over the process. Restrictive site-specific conditions as well as unfavorable soil conditions can be overcome in an above-ground system. In addition, the collection of flush water or elutriate is handled with far greater control than in a well extraction system. More significantly from a cost standpoint, the quantity of elutriate collected and, thus, to be treated, will be greatly reduced as natural aquifer water will not be pumped along with the flush water to the surface.

The process involves high energy contacting and mixing of excavated contaminated soils with an aqueous-based washing solution in a series of mobile washing units. The contaminated soil is first screened to remove rocks and other debris before entering a soil scrubbing unit, where it is sprayed with a washing fluid and subsequently rinsed. The next process is separation. The rinsed water or elutriate is recycled through a treatment system and reintroduced into the treatment process whereas the solids are settled out. Typically, the solids are separated into two groups: the fine-grained and coarse-grained fractions. Since contaminants are primarily concentrated in the fine-grained fraction, these materials (silts and clay) are treated further in another contact chamber for enhanced separation. The sand fraction of the soil usually requires only the initial rinsing treatment and is removed from the system, dewatered, and disposed of.

REFERENCES

American Chemical Society Committee on Environmental Improvement, *Analytical Chemistry*, Vol. 55, p. 2210, 1983.

American Water Work Association, Water Quality and Treatment: *A Handbook of Community Water Supplies, Fourth Edition*, McGraw-Hill, New York, 1990.

ASTM Publication, *Standard Guide for Risk-Based Corrective Action Applied at Petroleum Release Sites*, Designation E1793-95, 1995.

ASTM Publication, *Standard Practice for Description and Identification of Soils (Visual-Manual Procedure)*, Designation D2487-84, 1984.

Benitez, J., *Process Engineering and Design for Air Pollution Control*, Prentice-Hall, Englewood Cliffs, NJ, 1993.

Cole, M. G., *Assessment and Remediation of Petroleum Contaminated Sites*, CRC Press, Boca Raton, FL, 1994.

Calabrese, E. J. and P. T. Kostecki, *Principles and Practices for Petroleum Contaminated Soils*, Lewis Publishers, Boca Raton, FL, 1993.

Davis, S. N. and R. J. M. De Wiest, *Hydrogeology*, John Wiley & Sons, New York, 1966.

Driscoll, F. G., *Groundwater and Wells*, Johnson Filtration System, Inc., SE. Paul, MN, 1986.

Freeze, R. A. and J. A. Cherry, *Groundwater*, Prentice-Hall, Inc., Englewood Cliffs, NJ, 1979.

Heath, R. C., Basic Groundwater Hydrology, U.S. Geological Survey Water-Supply Paper 2220, 1982.

HWS Consulting Group, Inc., Site Characterization Report for the Garvey Elevator, Hastings, Nebraska, October 1995.

Johnson, P. C., C. C. Stanley, M. W. Kemblowski, D. L. Byers, and J. D. Colthard, A practical approach to the design, operation, and monitoring of in-situ soil venting systems, *Ground Water Monitoring Review*, p. 159, 1990.

Lawrence Livermore National Laboratory, *Recommendations to Improve the Cleanup Process for California's Leaking Underground Fuel Tanks (LUFTs)*, UCRL-AR-121762, 1995.

Lovley, D. R., M. J. Baedecker, D. J., Lonergan, I. M., Cozzarlli, E. J. P. Phillips, and D. L. Siegel, Oxidation of aromatic hydrocarbons coupled to microbial iron reduction, *Nature*, Vol. 339, 297, 1989.

Maidment, D., *Handbook of Hydrology*, McGraw-Hill, New York, 1993.

Mikesell, M. D., J. J. Kukor, and R. H. Olsen, Fundamentals of Bacterial Degradation of BTEX, *Proceedings of the CoBioRem Conference on Bioremediation of Petroleum Hydrocarbons in Soils and Groundwater*, Lansing, Michigan, March 1–2, 1993.

National Research Council, *In Situ Bioremediation: When Does it Work?*, National Academy of Sciences, Washington, DC, 1993.

Onda, K., H. Takeuchi, and Y. Okumoto, Mass transfer coefficients between gas and liquid phases in packed columns, *Journal of Chemical Engineering of Japan*, Vol. 1, p. 56, 1968.

Salanitro, J. P., The role of bioremediation in the management of aromatics hydrocarbon plumes in aquifer, *Groundwater Monitoring and Remediation*, Vol. 13, p. 150, 1993.

Sawyer, C. N., P. L. McCarty, and G. F. Parkin, *Chemistry for Environmental Engineering*, Fourth Edition, McGraw-Hill, New York, 1994.

Sherwood, T. K., and F. A. L. Holloway, *Transactions of American Institute of Chemical Engineers*, Vol. 36, p. 39, 1940.

Taylor, S. W. and P. R. Jaffe, Enhanced in situ biodegradation and aquifer permeability reduction, *Journal of Environmental Engineering*, Vol. 117, 1991, p. 25.

Taylor, S. W., P. C. D. Miller, and P. R. Jaffe, Biofilm growth and the related changes in the physical properties of a porous medium permeability, *Water Resources Research*, Vol. 26, p. 2161, 1990.

Thomas, J. M. and C. H. Ward, In situ bioremediation of organic contaminants in the subsurface, *Environmental Science and Technology*, Vol. 23, p. 760, 1989.

Tri-Regional Board Staff Recommendations, State of California, 1990.

Uniform Plumbing Code, Chapter 12: Fuel gas piping, *Uniform Plumbing Code*, 1994.

U.S. Environmental Protection Agency, *A Technology of Soil Vapor Extraction and Air Sparging*, by Loden, M. E., 1990a.

U.S. Environmental Protection Agency, *Air Pollution Engineering Manual*, by Danielson, J. A. (Editor), 1973.

U.S. Environmental Protection Agency, *Basic of Pump-and-Treat Ground-Water Remediation Technology*, EPA-600/8-90/003, 1990b.

U.S. Environmental Protection Agency, Guide for Conducting Treatability Studies Under CERCLA: Aerobic Biodegradation Remedy Screening — Interim Guidance, EPA/540/2-91/013A, July, 1991b.

U.S. Environmental Protection Agency, *In Situ Bioremediation of Groundwater and Geological Material: A Review of Technologies*, by Robert S. Kerr Environmental Research Laboratory, EPA/600/R-93/124, July, 1993.

U.S. Environmental Protection Agency, *Movement of Bacteria Through Soil and Aquifer Sand*, by Alexander, M., R. J. Wagenet, P. C. Baveye, J. T. Gannon, U. Mingelgrin, and Y. Tan, EPA/600/2-91/010, March, 1991a.

Water Environmental Federation, *Wastewater Biology: The Life Processes*, 1994.

INDEX

A

B

H

I

O

P

Q

R

S

T

U

V